CHARLES DICKENS'S NETWORKS

The same week in February 1836 that Charles Dickens was hired to write his first noval, *The Pickwick Papers*, the first railway line in London opened. *Charles Dickens's Networks* explores the rise of the global, high-speed passenger transport network in the nineteenth century and the indelible impact it made on Dickens's work. The advent first of stage coaches, then of railways and transoceanic steam ships made unprecedented round-trip journeys across once seemingly far distances seem ordinary and systematic. Time itself was changed. The Victorians overran the separate, local times kept in each town, establishing instead the synchronized, 'standard' time, which now ticks on our clocks. Jonathan H. Grossman examines the history of public transport's systematic networking of people and how this revolutionized perceptions of time, space, and community, and how the art form of the novel played a special role in synthesizing and understanding it. Focusing on a trio of road novels by Charles Dickens, he looks first at a key historical moment in the networked community's coming together, then at a subsequent recognition of its tragic limits, and, finally, at the construction of a revised view that expressed the precarious, limited omniscient perspective by which passengers came to imagine their journeying in the network.

CHARLES DICKENS'S NETWORKS

Public Transport and the Novel

JONATHAN H. GROSSMAN

OXFORD
UNIVERSITY PRESS

OXFORD

UNIVERSITY PRESS

Great Clarendon Street, Oxford, OX2 6DP,
United Kingdom

Oxford University Press is a department of the University of Oxford.
It furthers the University's objective of excellence in research, scholarship,
and education by publishing worldwide. Oxford is a registered trade mark of
Oxford University Press in the UK and in certain other countries

First published 2012
First published in paperback 2013

Impression: 1

Published in the United States of America by Oxford University Press
198 Madison Avenue, New York, NY 10016, United States of America

British Library Cataloguing in Publication Data
Data available

Library of Congress Control Number: 2011944133

ISBN 978–0–19–964419–3 (Hbk.)
ISBN 978–0–19–968216-4 (Pbk.)

Printed in Great Britain by
Ashford Colour Press Ltd., Gosport, Hampshire.

Acknowledgments

This book is about people's journeying. Not wanting the subject of my analysis to be confused with my analysis of it, I eschew creating metaphors myself about journeying, leaving others to suggest, for instance, that there is no frigate like a book to take one lands away (Emily Dickinson). Here, however, I want to make a single exception for an exceptional colleague and friend: Irene Tucker. Irene has been my fellow traveler and guide as I wrote this book. Throughout its writing, she frequently saw much better than I did where it should go and how to get there. (Whether I actually arrive is another matter.) All along the way, in a rare gift that only the truly gifted can give, Irene taught me as much as I could learn about thinking through my arguments.

Every year at the opening of the Dickens Universe summer conference in Santa Cruz, California, the faithful director, John Jordan, declares in a grand metaphor that Charles Dickens represents 'a railway station through which all things Victorian pass'. This book might help explain the aptness of Jordan's metaphor. This book's existence also partly depended upon this annual conference and the many experts—not all academics—that it assembled. I am especially grateful for valuable conversations I had there about my research with John Bowen, Andrew Miller, Bob Newsom, Bob Patten, Robyn Warhol, Carolyn Williams, and Alex Woloch.

I am also indebted to my intelligent and witty colleagues at UCLA, who create a stimulating and supportive environment in which to work. In particular, year in and year out, Helen Deutsch, Chris Looby, Yogita Goyal, Sianne Ngai, and Michael North provided wisdom and energy, professional and intellectual. Anne Mellor and Felicity Nussbaum cared about this book and my work, and that felt like a gift. A number of colleagues read or responded directly to my work in progress: Mark Seltzer quickened my thinking about systems; Mark McGurl demanded to see the big picture; Joseph Bristow was a resource for all things Victorian; Kirstie McClure, early on, suggested the fallacy of applying political discourse to interpret a transport-networked community; later, Saree Makdisi reminded me not to forget about politics completely. Tom Wortham was department chair when this project began and Ali Behdad when it was completed; both mentored and sponsored the research that appears here.

Many people, too many people to list, provided valuable information, judicious counsel, or other forms of assistance. To name just a few: Nina Auerbach, Julie Crawford, Ian Duncan, Matt Dubord, Jen Fleissner, Maria Frawley, Natalka Freeland, Dustin Friedman, Holly Furneaux, Philip Joseph, Richard Kaye, James Landau, Ron Lear, Andrew McNeillie, Richard Menke, Elsie Michie, John Plotz, Leah Price, Josie Richstad, Simon Stern, Gillian Silverman, Robert Thornton, Lindsay Waters, and Julian Yates. This book is better than it would have been thanks to Hilary Schor, who shared her deep understanding of Dickens, and Helena Michie, who, after reading chapters or listening to arguments, always responded with searching critical questions. My long-time mentor John Sutherland took time out to critique the first draft of the manuscript. As he gently reminded me, the book still needed an introduction, and my efforts to produce one benefited greatly from the fresh eyes of Jayne Lewis and Talia Schaffer. I also extend my heartfelt thanks to my three anonymous readers. As they will recognize, I pounced on virtually all of their many superb, deeply intelligent suggestions, even sometimes directly adopting their words and ideas. At Oxford University Press, the book gained an expert editor in Jacqueline Baker.

I recruited much help from friends and family as well. Two creative masterminds, Rayna Kalas and Liza Yukins, have kept me sane. Adam Parker deserves a prize for making me laugh and think at the same time. For broadening my perspectives, I thank Tim Mackey, Jane Penner, Etsu Taniguchi, Laura Wason, my sister Gillian Grossman, and my brother Nicholas Grossman. I will always be grateful to my mother Penny Grossman for the talks about this book over tea at the Life Boat House, Isle of Wight, and to my father Marc Grossman for pages of insightful comments on the chapters. Eli, my ten-year-old son, gave me much sanity-saving advice ('maybe you need to have fun with it?'), while my seven-year-old daughter Dhalia, whom I promised to get a dog after finishing my book, asked me almost every day the question that others began to feel they couldn't voice: 'Did you finish your book yet?' Her good-humored determination encouraged my own. My partner Jana Portnow helped me talk through all the ideas in this book and helped edit it too. How, though, to acknowledge her endlessly tested patience and daily support for the writing of this book? One day, years ago, when I was momentarily off somewhere else, Jana found my computer running, and she typed some words into one of the chapters. I stumbled across her little addition sometime later, and it made me smile. I cut and paste her words here; they are oh-so-true: 'JP is the best and I worship her! Love, JG.'

Contents

List of Illustrations

Introduction

In 1829, British colonist and colonized Indian alike were not sucked direct to Bengal by Grand Vacuum Tube boarded at Greenwich Hill. Nor, as William Heath also inventively pictured (Fig. 1), once there, did a short climb allow anyone access to a convenient company suspension bridge to South Africa, restaurants built into the towers. No one ever saw a solo aviatrix threading via kite through the sky, apparently steering with a modified spindle. Balloon platforms never airlifted troops; nor were any convicts transported to Australia by winged mechanized bat-monster. On the water, at no time did pilots take whip in hand to drive boats harnessed to sea creatures. On the highways, once crawling wagons, now motorized, did not do London to Bath in six hours. And no farmer, no gentleman, no squire, and no lady ever commuted by a steam-horse called Velocity on an express with 'no stop[p]age on the road'.

Yet such impossibilities pictured the real sense in the 1820s that a revolution in passenger transport was realizing impossibilities. This book is about this nineteenth-century revolution in passenger transport. It examines its actual—and not much less amazing—history.

It is a busy picture. In the engraving, partly the busyness is simply typical of the 'March of Intellect' series to which it belongs; the series' images intend to overwhelm its viewers with zany, futuristic possibilities in whatever realm they depict. Here the busyness also reflects, however, public transportation's busyness, evoking people everywhere in speeded motion across a shrinking globe. Its scattered disorganization mirrors a spatialized content in which ordinary people seem flying off in all directions, for all sorts of separate reasons. And, as I recreated in my opening's partial redescription, the viewer's attention largely hopscotches between different vehicles, each a self-contained, miniature specimen. The artist aims to portray a panoply of imaginary modes

Figure 1. William Heath, 'March of Intellect', London: T. McClean, 1829. Courtesy of the National Library of Australia.

of mobility, all discretely operating at once. (The tube connecting to the bridge represents an exception to the rule.) 'Vont you take a-hire Joe', barks a man in the center-front foreground, through a mouth stuffed with a pineapple, symbol of hospitality. While the black child servant, opposite him, propping up an umbrella, undercuts the intended implication that everyone falls equally under the umbrella of the question (which is also gendered—'Joe'—and addressed to the well-to-do), the question nonetheless accurately captions the picture as offering up to its viewers all its separate, futuristic means of passenger transport.

It cuts against this picture's grain to ponder the stories of the passengers. On the steam horse, a wigged squire may be leering at a woman, but generally, interaction, which might cue personal narrative, is noticeably absent. No one is greeting anyone; no one is waving goodbye. The artist does not render these travelers as coming and going from their meetings and partings. Nor does he survey their crisscrossing with strangers or their journeying in round-trip circuits, with, for instance, two of the same conveyances passing each other in opposite directions. So though they are all in one frame, the people hardly appear circulating in a single story. The only thing coherent would seem to be that their diverse world is shrinking, their means of mobility is accelerating, and everyone all around seems on the go (not in fact on a 'march' together). What else could that indicate but a mishmash of separate communities?—Buddha sits atop St Paul's cathedral dome; the ordinary man staffing the 'Royal Patent Boot Cleaning Engine' (lower right) is reading the *Gazette de Français*. It's geo-kinetic social chaos.

The picture falls short of imparting any sense that the advances in public transport were interconnecting passengers by networking them together. It envisions people separately zooming about, and it deduces that their separate communities mingle hodgepodge. More than merely being a byproduct of its technological focus, this inability to imagine passenger transport in networked terms calls attention to the fact that the networking of passengers had, historically, to be comprehended— and that it was perhaps even somewhat difficult to comprehend. People, after all, had always traveled. What was the difference? Weren't they just doing it more and faster? This book is about the history of public transport's systematic networking of people and the difference it makes—how it revolutionized perceptions of time and space, how it involved re-imagining community, and how the art form of the novel played a special role in synthesizing and understanding it. In a nutshell, this book looks first at a key historical moment in that networked community's coming together, then at a subsequent recognition of its tragic limits, and, finally, at the working out of a revised view that expressed the precarious, limited omniscient perspective by which passengers came to imagine their journeying in the network.

Today most people primarily associate the nineteenth-century public transport network with the birth of the railways. And because the railways laid down a brand new, wildly successful passenger transport system essentially from scratch, they were, in that way, revolutionary. In the history of passenger networks especially germane here, however, the railways were also continuing an acceleration and systematization notably brought together previously by stage coaching. Consider again, for instance, the engraving. It declares a passenger transport revolution is under way: where are the railways? There is no hint of them, though the picture appeared the very same year (1829) that steam locomotives on rails famously broke the speed barrier set by horse-drawn coaches at the Rainhill trials and just the year before the world's first successful passenger commercial railway began running between Liverpool and Manchester, spelling the doom of stage coaching. Instead, however, the only futuristic invention this artist correctly saw coming was intercity highway-traveling steam coaches. These did appear fleetingly on the roads before surrendering to the railways, and that steam horse's promise of 'No stop[p]age on the road' announces pretty precisely the advance steam might seem to offer in 1829 to *stage coaching*—obviating the long-distance coaches' requisite five-minute pit stops to change for fresh horses every ten miles. Similarly, the sole historical figure to whom the picture refers is not George Stephenson, the father of the railways, whose locomotive *Rocket* would win at Rainhill. Rather, the artist depicts (in the left background) a giant, wheeled, motorized, multispouted watering pot labeled 'McAdams'. This bizarre invention, designed to lay the dust notoriously kicked up by macadamized roads, alludes to John Louden McAdam, pioneering engineer in smoothing and waterproofing road surfaces, progeny of the eighteenth-century's new turnpike system for maintaining roads.

This book aims, in part, to recover the significance of the rise of a fast-driving, stage-coach network that systematized—before the railways—swift, circulating, round-trip inland journeying, with regular schedules, running continuously, available to ordinary passengers. The railways copied and intensified this system as a system, and, in this way, the railways will be seen to matter greatly here.

This nexus of stage coach and locomotive as part of a single transformation was somewhat more apparent in its own time. When, for instance, in 1866, in introducing *Felix Holt*, George Eliot wished to recollect the historical transitions under way just before the first reform bill in 1832, she evoked a stage-coach journey across the Midlands. For her, rendering a trip from the point of view of the coachman and the passenger outside on the box usefully traversed the changing national landscape. They pass, for instance, from country village to industrialized city, 'from one phase of English life to another'. More than that, though, the swift stage coach itself also represented an essential change. Some provincial people, Eliot imagined, might still not yet know how to interpret it. Racing past a shepherd, it seems part of a 'distant system of things called "Gover'ment"'; overtaking a rich farmer, it appears 'an accommodation for people who had not their own gigs, or who, wanting to travel to London and such distant places, belonged to the trading and less solid part of the nation'.[1] Thus did Eliot rework Thomas De Quincey's earlier, similar, powerful retrospective vision in *The English Mail Coach, or The Glory of Motion* (1849) of a nation becoming networked by stage coaches (more, briefly, on De Quincey in Chapter 1).

It was Charles Dickens, though, who most deeply understood that the stage coaches systematized public transport and the difference it made. De Quincey and Eliot are typical in translating passenger transport into some other system—in their case, national politics. Dickens, by contrast, repeatedly isolated passenger transport's networking effects and explored how the notionally mobile characters of his novels offered the means for comprehending the networking of passengers. And, rather than casting a wide net, this book focuses on Dickens. In fact, my argument, to restate it now more fully, is that Dickens realized the historic transformations wrought by stage coaching in his time initially by celebrating its unifying, communal dimension in *The Pickwick Papers* (1836–7), his first novel. Then, after the railways consolidated this system, the community it networked began to reveal more clearly its tragic dimensions, and, noticing this, Dickens rendered it in depicting the fatal intercity trek of little Nell in *The Personal Adventures of Master Humphrey: The Old Curiosity Shop* (1840–1). Later still, the system's international reach became increasingly apparent, and Dickens showed the implications of this in *Little Dorrit* (1855–7), with its stunning opening onto Marseilles and the activity of 'fellow travellers' abroad pictured as happening simultaneously. By that point also, however, neither the networked community's comic nor its tragic aspects seemed to count most, but rather the narrative perspective—which Dickens had all along been developing—capable of taking in its precarious formation.

This trajectory traces, through a trio of road novels (treated in three separate chapters), an evolving understanding of the passenger transport system. In each instance, Dickens takes the measure of the passenger transport revolution from a different present moment and returns to depict roughly the same period—the 1820s and 1830s—because each time he is discovering, along with his readers, something new that retrospectively can be seen to have always been true of the way the passenger transportation system networks people, warps space and time, and transforms the art of the novel, which provides a means for its comprehension. Though he too could see earlier evolutions in passenger transport, Dickens looked back over and again to changes that he associated primarily with the stage coaches, and he did so with an acuity that was something like the opposite of a hazy-dazy nostalgia: shedding light for his readers on how what had been wrought then was making their present.

Along with his readers, Dickens comprehended their networking by public transport through narrative, and their networking by public transport entailed reconfiguring narrative form, particularly the picaresque road novel and travel literature. In Dickens's hands, the novel as an art not only could enable his community, whose individuals were increasingly atomized, to come to know their manifold unseen connectedness, but also, more specifically, could help to produce its self-comprehension in terms of a crisscrossing journeying of characters simultaneously circulating all around. (This is just what the artist of the engraving missed.) As Chapter 1 explains, *Pickwick* helped craft a recognition of this networked community structured around a 'simultaneous plurality', to quote J. Hillis Miller's apt phrase. Dickens had, however, silently supplied the omniscient narratorial view that conjured the simultaneous, eclectic activity happening all around a passenger transport network. As Chapter 2 explores, Dickens recognized in *The Old Curiosity Shop* some of the tragic limits of the networked community, and that meant also confronting the construction of that supervising omniscient perspective. In a multiplotted novel, the omniscient narrator had to assemble retrospectively and serially the characters' simultaneous circulation; a coordinating 'Meantime...' logic bound together community. *Little Dorrit*, the subject of Chapter 3, would subsequently represent the warping effect of stretching that 'Meantime...' across an international terrain. It also again presented a totalizing omniscient perspective of a community figured as fellow travelers, 'journeying by land and journeying by sea, coming and going so strangely, to meet and to act and react on one another' such that 'in our course through life we shall meet the people who are coming to meet *us*, from many strange places and by many strange roads'.[2] Now, however, Dickens fashioned a story that exposed that individuals, who grasped their networking through projecting themselves from a third-person perspective, could never actually achieve a comprehensive picture of their plottable plots. Only belatedly and dimly could they glimpse their crisscrossing, closely networked relations with others.

As this outline indicates, certain narratological complexities form a substratum of this book: especially omniscient narration, simultaneity, serialization, and multiplottedness. At its most general level, though, this is a book about imagining community in an era of systems and networks. That term 'community' will necessarily change in meaning here. It is what is getting redefined, and a clear perception that its definition is perplexing and unstable—sometimes, for instance, awkwardly provoking its replacement by the less positively inflected term 'collective' and other times making its modification by the adjective 'networked' feel virtually redundant—is fully part of the point. Meanwhile, no sharp terminological distinction is made here between the term 'network', which tends to emphasize connectivity and its extensions, and the term 'system', which tends toward enclosing—if ever shifting—boundaries and self-defining processes. While sustaining those differing inflections, both aspects are treated as jointly in play.

This book does sharply distinguish, however, the passenger transportation system, which delimits its subject, from the communication system, which only appears here in distinction from the passenger transport system. (In this book, 'communicating' never waffles into its secondary sense of indicating a physical passage between two places.) The differences between those two affiliated systems are often, and sometimes quite correctly, blurred, collapsed, or braided, but this book aims to orient its readers to a separate history of passenger networks. Hence, for example, in cropping slightly the engraving of futuristic passenger transport (Fig. 1), I intentionally pruned out a winged postman, who appeared in the lower left corner. Thus—as that which is chopped out—does he re-enter here. As especially the conclusion of Chapter 1 will show, Dickens hones distinctions between the communication system, to which medium his novels belong, and the passenger transport system, about which they have so much to say.

As my subtitle, *Public Transport and the Novel*, announces, this book thus stakes out a different angle than the mass of critical attention paid in the past few decades to communication systems. Taking up the intellectual mantle of Marshall McLuhan ('the medium is the message'), Friedrich Kittler—beginning with his *Discourse Networks 1800/1900* (1985, trans. 1990)—now towers behind a diverse, interdisciplinary discussion of media. A couple of relevant, recent direct descendants are Bernhard Siegert's *Relays: Literature as an Epoch of the Postal System* (1993, trans. 1999) and Cornelia Vismann's *Files: Law and*

Media Technology (2000, trans. 2008). For Victorianists, there is especially Richard Menke's *Telegraphic Realism: Victorian Fiction and Other
Information Systems* (2008). Menke's introduction to the field need not
be recapitulated here; rather, precisely because so many excellent
books have sharpened and reshaped our sense of the communication
revolution (which continues to rage all around us today), we need a
book that does the same for passenger transport. This book aims to
use networking to rethink the nineteenth-century revolution in passenger transport in the same way that Kittler, Menke, and others have
used it to rethink Victorian talk. It is about Brunel, not Babbage, cabs,
and not copyrights.

Separately from communication systems, but in conjunction with
them, public transport occupies a similarly second-order, infrastruc-
tural relation to other systems. In organizing the circulation of living
bodies, it refines the form of other systems—economic, political, religious, and so on, which all compel people's circulation—into its contents. In the nineteenth century, transport most obviously helped to
fulfill the activities of commerce and government—e.g. 'Direct to
Bengal'—and this book will seem to deflate the historical importance
to passenger transport of commerce, government, and such other systems. This is a fault to which it accedes because it aims to clear some
ground to sharpen understanding other systems in relation to passenger transport.

Consequently, this book draws together the work of transport historians.[3] Especially informative has been Philip Bagwell's *The Transport
Revolution* (1974), and I have intentionally adopted Bagwell's title as
something of a refrain in this book. This is perhaps somewhat incautious.
Bagwell wisely spreads his transport revolution across the entire eighteenth century through to his late twentieth-century present, and this
study focuses much more narrowly. I intend, however, the term 'revolution' only to refer back to Bagwell's broader, uncontroversial sweep, not
to bog readers down in assessing the comparative impact of one historical advance over another or whether, at some threshold, acceleration
equals revolution. This book's cultural and theoretical approach to transport history has been inspired by Wolfgang Schivelbusch's *The Railway
Journey* (1977). Schivelbusch enters creatively into details of transport
history to call attention to its transformative social experiences.

As I mentioned earlier, I especially scrutinize the novel's multiplotted 'Meantime...' structure by which it renders the simultaneous

activity of a community as well as marking that community's limits. My starting point here is Benedict Anderson's *Imagined Communities* (1983). Anderson brilliantly argued that the novel's multiplotted structure abets the imagining of a nation as the simultaneous activity of its citizens marching through history. I explore here a differently imagined community, but one for whom imagining simultaneity also matters. In this community's history, the rise of an 'empty, homogenous time', which enables national simultaneity in Anderson's argument, looks not so empty or homogenous—it is first called 'railway time', then standardized time.

A core story that this book will relay concerns how the systematic co-location of individuals' journeying standardized time and space—and not just in timetables and route maps. (Hence the first chapter's initial sections are entitled 'Time' and 'Space'.)[4] The picture of futuristic transport, with which this introduction began, foregrounds the spatial aspect, but the artist also gestures in a telling detail toward how the revolution in public transport warps time. He redraws St Paul's tower clock as sitting atop its cathedral dome and ticking inside the tummy of a Buddha. To this artist, that recombination likely represented something like both the erection of time as a new religion and the crazy collapsing of London time across separating physical and cultural distances. As this book will show, he was not so far off. It just took a Dickens to show that the passenger transport revolution was much more coherent than he realized.

I

The Speeding of the
Pickwick Coach

I. Time

Charles Dickens's first novel, *The Posthumous Papers of the Pickwick Club*, written and published in 1836–7, is about the late 1820s. Dickens stamps his story's date on the very first page. Assuming the guise of an editor, the narrator declares that the 'first ray of light' that converts into 'a dazzling brilliancy...the earlier history of the immortal Pickwick' derives from the 'perusal of the following entry in the transactions of the Pickwick Club' dated 'May 12, 1827'.[1] May 12, 1827 dates both the first of the Pickwick Papers and the meeting that originated the Pickwick Club's traveling branch, whose 'perambulations' follow, loosely occupying the next seventeen months in story time. The second chapter's opening sentence re-enforces the dating: 'That punctual servant of all work, the sun, had just risen, and begun to strike a light on the morning of the thirteenth of May, one thousand eight hundred and twenty-seven, when Mr. Pickwick burst like another sun from his slumbers' (1.5). Now Dickens writes the date out in prose, incorporating it into a description of the rising sun, which forms part of a past action just completed, after which Pickwick, bursting forth like another sun, begins the story anew as a character inhabiting his past present moment so fully that he becomes an emblem of time itself. Already these beginnings—that first light found in the archives, followed by the rising of the punctual sun, and then Pickwick bursting forth like another sun—put Dickens's 1836 readers into multiple relations to this story of the late 1820s. The first beginning is comically historiographical; the next reports direct from the past contemporary moment; and the last offers up a character constituted by his times.

This book originated in an attempt to understand *The Pickwick Papers'* relationship to time, in particular the way this novel seemed to link history, the present, and people's sense of their historical contemporaneity. Its impetus came from a simple question about the novel's temporal structure arising on the novel's opening page: why would Dickens, writing in 1836–7, so pointedly demarcate for his readers that this story occurred nine years earlier, in the late 1820s? Answering that question turned out to shed light on a central aspect of *Pickwick* that has preoccupied its critics and readers: the novel's renowned capacity to evoke (in Steven Marcus's oft-quoted phrase) 'the ideal possibilities of human relations in community'.[2] It also opened up the facet of nineteenth-century history that forms the subject of this book and, reciprocally, helped reveal that one of Dickens's achievements was to grasp the transformations involved in that history.

Before I begin to unravel precisely why *Pickwick's* focusing on a moment nine years earlier matters so much, I want to make it clear that despite his initial calendric specificity—May 12, 1827—Dickens has no strict investment in the actual recorded happenings of 1827. For instance, there is no real problem admitting anachronisms like the omnibus, only first introduced in London in 1829, if doing so makes for a good joke. I will later briefly compare *The Posthumous Papers of the Pickwick Club* to Walter Scott's breakthrough historical novel *Waverley; Or, 'Tis Sixty Years Since* (1814), but Dickens nowhere cares, as Scott certainly did, to interweave his fiction with the specific historic personages or political events of the years in which his story's action occurs.

Still, *Pickwick* is patently about the late 1820s, and we particularly know this because Pickwick—who takes his name from a stage-coach line—whirls along via stage-coach rides, introducing us to stage-coach drivers, coaching inns, and virtually every aspect of the stage-coaching system; yet a mere nine years later, in 1836–7, when Dickens was writing the novel, the steam railways had arrived and begun obliterating this world of long-distance stage coaching. This retrospective evocation of stage coaching during the onset of railway travel is a key to understanding *Pickwick*. Written and published as the railways were beginning to overtake long-distance coaching and to introduce an almost unthinkable velocity of movement, *Pickwick* transformed the recent era of stage coaching, newly understood in relation to the railway's dawning 'annihilation of time and space', into the novel's own creative principle of temporal and spatial coherence.[3] It projected the public transport system's essential aim—the coordinating of people's journeys in space and time—into a collective vision of individuals as synchronically engaged in interconnecting journeys and a model—across history—for community in contemporaneity.

It has been noticed before that *Pickwick* depicts a moment just before the beginning of passenger railway travel. James Kinsley, for instance, mentions it in his introduction to the Clarendon edition, and he—articulating the currently prevailing view—thinks the book thus manifests 'a calculated nostalgia'.[4] In this view, the railways metonymically represent modern industry, the coaches, a pre-industrial past, and the novel's celebratory evocation of coaching, nostalgia for that past. *Pickwick* thus becomes a last hurrah for Olde England and the simple country days of rambling coaches and cozy coaching inns. The trouble with this view is that anyone who has read *Pickwick* knows that it is not some slow novel of sleepy days. Just as it is always perpetually beginning again, it is always skimming along, racing, crashing, going—truly a novel of the road.

This speed especially belongs to the stage coaching of the 1820s and early 1830s. Dickens, born in 1812, grew up in a world where coaching was modernizing and accelerating, where he could discern both the residues of an older, displaced coaching time and a fast-driving system advancing so quickly as to outdate that which had been new just years earlier. He began working as a newspaper reporter for the *Morning Chronicle* in 1834, covering events outside of London from Edinburgh to Liverpool, from Bury St Edmunds to Birmingham, in trips that put him right on the cutting edge of stage coaching and made possible the later imagining of Pickwick's journeys. 'There never was', Dickens later wrote to John Forster in 1845, 'anybody connected with newspapers who, in the same space and time, had so much express and post-chaise experience as I.'[5] The pace was a race. Here is Dickens in an 1835 letter recounting one such race to deliver his report ahead of the London *Times*:

Dear Tom. I arrived here (57 miles from Exeter) at 8 yesterday Evening having finished my whack [his part of the news report] at the previous stage....I have now, not the slightest doubt (God willing) of the success of our Express. On

our first stage we had very poor horses. At the termination of the second, The
Times and I changed Horses together; they had the start two or three minutes:
I bribed the post boys tremendously & we came in literally neck and neck—
the most beautiful sight I ever saw. The next stage, your humble, caught them
before they had changed; & the next [the Times reporter] Denison preceded
Unwin about two minutes.... The roads were *extremely heavy*, & as *they* had 4
[horses], I ordered the same at every stage...[6]

As this letter gallops along, weaving a pressurized time sliced into minutes against the prize of delivering the news first, it makes plain that
the early nineteenth-century coaching network Dickens knew had
become as fast and time-sensitive as the newspapers whose immediacy
the coaches are here enabling. Racing coach against coach, neck and
neck, the most beautiful sight he ever saw, Dickens presses—he is
pressed—to match that other coach's speed. As a result, the individual
control one might naturally assume Dickens has over the coach—go
faster, go slower—diminishes. He has hired a post chaise (and so, for
instance, can lay on extra horses) rather than travel by a public stage
coach, on which he would have even less control, but still all he can do
is bribe those post boys to ride harder, faster, keep up. In a sense Dickens's post coach is bound to that other speeding coach, and in that
coupling as well as the resulting attenuation of Dickens's control, one
may glimpse the coming of the railways in which the coaches would
be literally coupled and passengers would surrender much of their
individual control. As one early treatise on the railways rued, a passenger 'has nothing to do but to put his head out of the coach window
and make his wants known; the coach can be stopped...[while] in
railway travelling.... [he] could by no possibility receive the least help'.[7]
And yet, as each of Dickens's breathless, first-person, active sentences
also trumpets, such an unstoppable public transport system could
equally empower its passengers, thrilled to leverage such a powerfully
organized network and delighted to blur their own bodies into its
system's space-and-time-warping swiftness.

Along with this expansive experience of coaching's acceleration, the
headlong pace of Dickens's letter also recalls his time-sensitized experience as a daily journalist. One should not be surprised that the news-
making event—the subject of his 'whack'—is not even worth
mentioning. Dickens the reporter is faithfully rushing here to satisfy
the demand for the news' timeliness itself, whatever the news. His
breakneck ride serves and connects a mass-reading public who were,

as Paul Virilio observes, 'not so much buying daily news as they were buying instantaneity, ubiquity—in other words, their own participation in universal contemporaneity'.[8]

It has long been understood that this kind of ongoing consumption by a community reading concurrently and to-the-moment production carried over, albeit with meaningful differences, in the serial publication of *Pickwick*, whose parts published on-the-go from the scribbling author each month were produced at an often desperate rate and always just to deadline. As Katherine Chittick, expert historian of *Pickwick*'s reception, uncovered, some reviewers even greeted the novel as a new periodical. And, as Kathleen Tillotson and John Butt long ago calculated in their ground-breaking *Dickens at Work*, Dickens gradually synchronized the timing of his story's events to report the passing months of his readers.[9] In the January number, for instance, readers heard all about the Pickwickians' glorious winter glide on the Muggleton Telegraph—'wheels skim over the hard and frosty ground.... speed[ing], at a smart gallop, the horses tossing their heads and rattling the harness as if in exhilaration at the rapidity of the motion' (10.284)—and then, in the story's real production of nationalist nostalgia for mythic Olde England, about their Christmas adventures at Dingley Dell.[10] By the final November number the serial readers could confidently expect to hear of Pickwick's October. Dickens firmly subjoined the entirety of this story to its monthly serial appearance.

What has not yet really been grasped is that Dickens thereby yoked his story's depiction of 1820s coaching to his readers' current 1830s calendars. This confluence meant that the first readers of *Pickwick* experienced the stage-coaching adventures that occurred roughly a decade ago to Pickwick, his manservant Sam Weller, and three Pickwickian followers, with a newspaperish sense of contemporaneity. This road novel, with its five journeys, reported on a past present-time speeding exuberantly along with four horses laid on at every stage. Present thus met past in this novel in a conjunction that the word 'contemporary' itself acquired in the Victorian period as it came to mean not only 'of a certain shared time', but also, as in more common parlance, 'of the present time', 'modern', and 'now' (as in 'contemporary art').

In doing so, *Pickwick* returned, I am suggesting, to the late 1820s and the last days of stage coaching not to eulogize their passing, but to think through the almost-unthinkable further speeding up that the railways were introducing in the 1830s, when it was written. Dickens portrays, and revels in, not some bygone era of stage coaching but the speeding up of the stage coaches that had preceded, and was still occurring in, the period he describes. This is not to deny that, just like Pickwick's renowned black tights and gaiters, the novel's stage coaching would feel dated to its 1830s readers. It is to say that this datedness does not mean that Dickens is looking back nostalgically. Rather, much as Pickwick's tights and gaiters are meaningfully already out-of-fashion within the story, the novel's stage coaching thrusts its readers into something like the temporal experience of the rapid passing of fashion itself.

Looking back less than a decade, Dickens identifies a people caught up in an acceleration of change and telescoping of time in which they could see and experience themselves as rapidly becoming dated. In doing so, the novel does not return to the 1820s to showcase an early piece of industrial history akin to its own newly emerging railway-based present. *Pickwick* is not a historical novel drawing a parallel between a present age and a past one. Rather, this novel retrospectively represents, with an exhilarating newspaper-like form, an earlier, previous 'generation' that came to be defined technologically both by the passenger transport network's acceleration and by the resulting communal experience of that generation's own present as a passing historical moment. But instead of representing this passing-away as a threat to the coherence of that shared world, the achievement of *Pickwick* was to transform a modern awareness of one's times as fleeting, passing, always-in-the-process-of-becoming-what-was, into a sense of community.

To understand this effect, it is necessary to link the history of two time-shaping technologies that are usually contrasted: that of the stage coach and of the steam locomotive. These two passenger transport technologies together made for much of society's ever-quickening pace in the first half of the nineteenth century, and a picture from the 1823 edition of Thomas Gray's *Observations on a General Iron Rail-way* illustrates their intersection in Dickens's young adulthood (Fig. 2).

In *Observations on a General Iron Rail-way*, Gray is imagining, and lobbying hard for, the building of a national passenger railway system. His audience likely would have been struck by his illustration's futuristic use of a steam engine. He has ripped that engine right out of its coal-mining habitat. (Hence it still has a giant cog underneath, which, as railway historian and theorist Wolfgang Schivelbusch explains, meshes both conceptually and literally with the unified machine being

Figure 2. Train of Coaches. Frontispiece (detail) from Thomas Gray, *Observations on a General Iron Rail-way* (London: Baldwin, 1823). Courtesy of Rare Books and Manuscripts Department, The Sheridan Libraries, Johns Hopkins University.

used to draw coal from mines but jars with the later conceptualization of the locomotive as a self-contained conveyance that rides the rails where it will.)[11] Likely most striking for readers today is, however, that Gray's futuristic conception of passenger rail travel couples old-fashioned stage coaches—their drivers and bugle guards included—to a locomotive. A modern, periodizing eye typically partitions stage coaching from railway. In reality, stage coaching had advanced so radically in the years preceding the invention of the steam railroad that by the 1820s, when Gray imagined this machine, his proposal for a national passenger steam-powered rail system naturally envisioned it not as a replacement for the stage-coaching system but as its logical next step.

Two advances in coaching technology especially converged in the early nineteenth century to produce a new fast-driving system.[12] One major advance was the full incorporation of metal springs into coach construction, providing the carriages with their first truly effective suspension systems. A basic challenge that all coaching faced throughout its early history was that the ride was a nightmare of jarring jolts. In a sketch published in 1835 in *The Evening Chronicle*, and later reprinted in *Sketches by 'Boz'* (1836), entitled 'Early Coaches' (and that title itself partly makes the point), Dickens sarcastically declares: 'we should very much like to know how many months of constant travelling in a succession of early coaches an unfortunate mortal could endure. Breaking a man alive upon the wheel would be nothing to breaking his rest, his peace, his heart—every thing but his fast—upon four.'[13] As this vivid comparison suggests (and one may suspect that breaking fast here alludes not merely to starving but to vomiting en route), some early coaches still survived to torture passengers in the early nineteenth century. By then, however, advances in suspension systems had transformed and accelerated the ride. A good suspension system meant coaches could go fast without pulping their passengers. It also allowed a wheel to climb lightly in and out of ruts without having always to lumber under the coach's weight. Because the rest of the coach would then not have to endure nearly as much shaking, the coach could also be made of much lighter materials, and lightening the coach meant added speed.

The other major breakthrough in coaching technology, and the one to which historians of coaching invariably point as probably the single most important reason why 'the golden age of coaching', as they sometimes call it, dawned in the early nineteenth century, was that the network of ruts with sometimes four-foot-deep holes, called the highways, had been vastly improved.[14] Up until the eighteenth and nineteenth centuries, first monasteries, then Christian parishes, had mostly failed to build or sustain intercity roads, which, by and large, mattered less to them than to a few infrequent strangers passing through. This earlier era might not be entirely unfairly epitomized by an image of Norman conquerors building their stone castles from pavement ripped from Roman roads. Road historian Graham West synoptically titles the entire millennium following the Romans the 'Years of Neglect'. State-sponsored protective restrictions of the roadbeds were the order of the day. Again and again, the government—primarily oriented toward the trucking of goods, not passengers—struggled in different ways to shield pathways from destruction by thin, sharp cutting wheels and ground-churning heavy loads. During these many centuries—the sixth to the sixteenth—'road' did not so much describe a built part of the landscape as a common legal right-of-way across it. Over the miry, rutted routes, sledges might honestly compete with wheels. Walking was the common lot, and it was typically as fast as any vehicle: the chariot, the whirlicote (i.e. a cot on wheels), the common wagon— nothing called a coach existing till the 1600s. Riding horseback, a mode of transport never much bound to the road, remained the fastest means of inland mobility.

Then, in the eighteenth century, a new turnpike system—an administrative scheme of toll gates—began financing the maintenance of roads by those who traveled them.[15] The turnpike system gave birth to the first professional road engineers, including two early heroes of transport history, Thomas Telford (1757–1834) and John Louden

McAdam (1756–1836). These two men quite literally laid the groundwork for the revolution in passenger transport. Telford was called 'the
colossus of the roads'. He famously constructed thick, rigid road foun-
dations; these were necessary across slipshod pathways, but came at a
high cost and required much labor. He also formed his road surfaces
with round rocks, which made for more bumps, worse potholes, and
lame horses. In the decades preceding the 1820s of *Pickwick*, John Lou-
don McAdam perfected a new successful surfacing technique—later
called macadamizing. McAdam drained the route and spread it with
angular, broken pieces of hard stones gauged in size so that under
wheels and hooves they would flatten together, providing a smooth
and waterproof surface (but not as hard on or as slippery to hooves as
the clattering granite paving of city streets). The straight, single line
that Thomas Gray draws beneath the coaches pulled by the locomotive (in Fig. 2) is meant to represent the rails, but from the perspective
of coaching technology that solid line and the rails themselves represented part of an ongoing revolution in remaking the road as a fric-
tionless but traction-giving utopia for the wheel. Smooth roads, light
coaches: by the 1820s, for the first time ever, the limits of coaching
technology approached the speed and exhaustibility of the horses that
were the coach's motor—until the iron-horses took over.

Much more could be said about the technological advancements of
stage coaching. There was, for instance, the stage coaches' dished
wheels, each component of which required different types of wood of
differing flexibility—elm, oak, ash—with iron rims. Or, there was the
vast industry of horse breeding through which the coach's horse would
become, in the description of an 1831 treatise, 'as different from what
he was fifty years ago as it is possible to conceive'.[16] My aim here,
however, is simply to indicate this technological advancement and to
make real a conjunction between stage coach and railway that may
seem initially odd.

Long-distance stage coaches sped up. Philip Bagwell offers this
handy general measure: 'Comparing the 1750s with the 1830s journey
times on the main routes linking principle cities were reduced by
four-fifths; comparing the 1770s with the 1830s the times were halved.'[17]
Dickens routinely registers the accelerating journey times in *Pickwick*—Bath from London, having been for most of stage-coach history
three very hard days, shaved to two days in living memory, has become
by the late 1820s a well-regulated day trip. 'Mind—the seven-mile

stage in less than half an hour!' Wardle will shout (4.85), relying on pushing his hired post chaise to fifteen miles per hour, one estimate of coaching's highest velocity.[18] However slow such a pace may now seem when gauged against planes, trains, and automobiles, to its contemporaries it truly represented an unprecedented acceleration.

And this unprecedented swiftness was vested in a public transport *system*. As important to passenger transport as any of the purely engineering advances was the reform of the postal service in 1784 by John Palmer. Palmer shifted the distribution of the mail from individually mounted post boys with letter bags to a speedier system of delivery by Royal Mail coaches that carried not only the post, but also a few paying passengers. Palmer's reform not only eventually helped ignite competition in the world of commercial travel by stage coach, but also, by yoking commercial passenger stage coaching to the postal system, it helped consolidate the long-standing public stage-coaching system as a system.

The fast-driving stage-coach system of Dickens's day was efficient, regular, interlinked, and continuously available, offering a polished national system of routes, with coaches leaving continually out of London for all the popular destinations (see Fig. 3), which were themselves linked along cross roads. The system ran on fresh horses, galloping back and forth between stages, available typically every ten miles at roadside inns, the whole network running night and day. Predictably, its drivers professionalized. By the time Dickens wrote *Pickwick*, the coachman was no longer commonly characterized as a surly, boozing plodder. He was a cosmopolitan, caffeine-drinking expert; slick in his rubberized Mackintosh (patent: 1823), he was as newly waterproofed as the roads; and, pictured with his vehicle fully loaded, discoursing casually, whip relaxed with 'four in hand', racing along the highway, he became an iconic image for Dickens, repeatedly illustrated by Phiz. He also, paradoxically, signaled the dawning ascendency of the passenger.

'By about 1820 it was quicker to travel by fast coach than on horseback,' calculates Bagwell; 'from this date the individual rider was much less frequently seen on the roads.'[19] By one newspaper's count, in 1830 'a person has 1,500 opportunities of leaving London in the course of twenty-four hours by stage coaches'.[20] Three dismaying, parenthetical words, previously ubiquitous, disappeared from the announcements of departure times: '(if God permit)'.

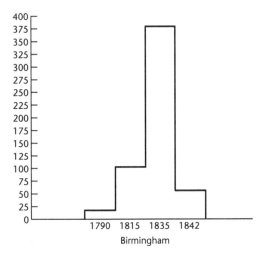

Figure 3. London to Birmingham, 'Development of Stage-Coach Travel 1790–1850: Number of daily departures of Royal Mail and stage coaches'. Reproduced by permission from Philip Bagwell, *The Transport Revolution* (London: Routledge, 1988), 32. © Taylor & Francis.

In the 1830s, the railways were replacing stage-coach travel, but not, as one might assume, because stage coaching represented the remnants and slower pace of an earlier agricultural era. With an estimated 700 Royal Mail coaches, 3,300 stage coaches, and 150,000 horses running a clockwork schedule, the early nineteenth-century's fast-driving coaching network represented to its contemporaries technological progress and accelerating life. It established the internal passenger transport system that the railways would copy. This is why it makes sense to say that *Pickwick*, depicting the 1820s, is not about the old coaching days, which would have roughly indicated pre-1790s to contemporary readers. In their experience, stage coaching had transformed before their eyes into a network so dense, routinized, and integrated that mobility within this system became a norm. In their experience, the stage coach's velocity had increased so much that the earlier days of coaching represented for them as much a decelerated, distant pre-industrial world as stage coaching itself would come to represent after the railways. What the steam-powered railways were conceptually to the 1830s and 1840s, modernized stage coaching was to the 1810s and 1820s.

By the time Dickens was writing *Pickwick*, steam-driven passenger railway travel was an expanding reality.[21] Its history is well known. In 1825, just a few years after Thomas Gray imagined his picture, the first truly operative steam-train line, the Stockton–Darlington, opened. Upon opening, it ran a locomotive pulling thirty-six coal carts and a single passenger coach, for visiting observers, called the *Experiment*, which looks essentially like a mini version of the coaches Gray envisions. Unless one wanted to commute on a coal-shipping route, however, there was as yet no such thing as passenger rail travel. Nor was there reason for it—steam locomotives were still slower than stage coaches. Then, at the Rainhill Trials in October of 1829, George Stephenson's new locomotive, *The Rocket*, triumphantly raced past the speed barrier of coaches, clocking in at speeds reportedly as high as thirty miles per hour, double coaching's fabled top speed. Within a year—in September 1830—the highly successful Liverpool–Manchester line opened. It was largely planning to carry freight. Unexpectedly, however, passengers flocked to it, and passenger rail travel began in earnest, with 'carriages . . . running regularly between the two towns'.[22]

This is why it makes sense to say that *Pickwick*, whose plot runs from 1827 to late 1828, depicts the final days of a fast-driving, long-distance stage-coach system that the steam-driven passenger railways were taking over. Initially, steam locomotion was even mixed in with horse-drawn towing, which had long been used in conjunction with rails. Similarly, the railed roads experimented with turnpike tolls, while, in a practice that would persist for years, which also conjoined the two modes of long-distance transport, private post coaches were mounted onto undercarriages for rail transport, their passengers riding inside, as Dombey does.[23] Neither did the coming of the steam railways mean that horse-drawn conveyances became obsolete, as is sometimes erroneously believed. With the number of circulating passengers exploding exponentially, horse-drawn transport became increasingly necessary and available—the omnibus, introduced in London in 1829, is a prime example. But on the routes competing with the steam railways, the stage coaches, though they fought back hard, were soon paying their last pike.[24]

On 10 February 1836, just two days after the first railway station in London opened, Chapman and Hall proposed to Dickens that he write the text for a monthly publication with illustrations by Robert Seymour. The stage-coaching novel almost immediately under way can productively be read against the ongoing construction of railways. And nowhere is this railway present clearer than in Pickwick's final and longest journey to the city of Birmingham. As Dickens wrote up this adventure (beginning with the seventeenth number, published at the end of August 1837), Birmingham represented the rail connection dramatically linking four of England's chief cities. In the north, the Grand Junction railroad had just completed the historic link of Birmingham to the Liverpool–Manchester line, while the first section of the line that would connect London up to Birmingham had opened.

In number seventeen, Dickens extricates Pickwick from Fleet prison, where he has been stewing for some time, and starts him on a brand new journey to Birmingham. Pickwick's loop first takes him west to Bristol, along a stage-coach route that many of Dickens's readers would have been well aware was doomed by the Great Western Railway, currently laying down its tracks. The number ends in the travelers' room of a Bristol coaching inn, 'The Bush' (out of which the real Bath coach proprietor Moses Pickwick operated coaches),[25] and there a bagman— that is, a traveling salesman—tells Pickwick the remarkable 'Story of the Bagman's Uncle', the novel's final interpolated tale. In this story, the bagman's uncle, Jack Martin, 'being very fond of coaches, old, young, or middle-aged', steals into an enclosure filled with 'old worn-out mail coaches' (17.521), 'the decaying skeletons of departed mails' (17.522). Here, initially, the broken array of junked coaches seems to present an irrecoverable past. Jack Martin sits 'himself quietly down on an old axletree...to contemplate the...mail coaches' (17.521) and falls asleep thinking of 'the busy bustling people who had rattled about, years before, in the old coaches' (17.522). Suddenly he awakes into a 'scene of the most extraordinary life and animation' in which the old mail coaches have sprung back to life: 'passengers arrived, portmanteaus were handed up, horses were put to, and in short it was perfectly clear that every mail there was to be off directly' (17.522).

Jack Martin is swept back into the first days of the mail coaches— the late eighteenth century and already accelerating stage coaching. He initially tries to resist the backward pull into that time: 'A mail travelling at the rate of six miles and a half an hour.... This shall be made known; I'll write to the papers' (17.526). But time overruns him. He finds himself in the midst of a mock, stock eighteenth-century romance tale of chivalry and betrayal, complete with sword-wielding, aristocratic villains kidnapping a beautiful young lady, all taking place on his coach journey. He battles the villains, commandeers the coach, and flees with his heroine in a hair-raising pursuit 'at fifteen good English miles an hour...whew! how they tore along!...faster! faster!....houses, gates, churches, haystacks, objects of every kind they shot by, with a velocity and noise like roaring waters suddenly let loose.... "faster! faster!"' (17.530)—until he awakes abruptly from this rushing dream-vision of the ghosts of the old mails.

Pickwick's last overdetermined interpolated tale thus presents both a *mise en abîme* of the novel and (like its northern setting in Edinburgh) a limit case. In it, Dickens tunnels with his readers from his story of the late 1820s and coaching life into a story told about an imagined return—relayed within appropriately dated generic narrative codes—to even earlier days of coaching. Both stories dramatize the recognition that the people of earlier times remain forever bound together in their present-moment narrative, while both also tie this periodizing, historicizing recognition to an era they together delimit by an accelerating public coaching network, one that serves not just aristocrats but ordinary people, Jack Martins.

Having condensed in this dream-tale this aspect of his novel's form, Dickens proceeds in the next number to relate 'How Mr. Pickwick sped upon his Mission' to Birmingham. Pickwick's ride is a showcase of the pleasures of 1820s stage coaching, but I want to pick up at his journey's darker end as the novel's stage coach rolls into the steam-driven world that is, for his readers, in the process of displacing that stage-coaching world:

It was quite dark when Mr. Pickwick roused himself sufficiently to look out of the window...the paths of cinder and brick dust, the deep red glow of furnace fires in the distance, the volumes of dense smoke issuing heavily forth from high toppling chimneys, blackening and obscuring every thing around...the ponderous waggons which toiled along the road, laden with clashing rods of iron...all betokened their rapid approach to the great working town of Birmingham. As they rattled through the narrow thoroughfares leading to the heart of the turmoil...The streets were thronged with working-people. The hum of labour resounded from every house...and the whirl of wheels and noise of machinery shook the trembling walls.... The din of hammers, the rushing of steam, and the dead heavy clanking of the engines, was the harsh music which arose from every quarter. (18.536)

Here the coaching of the 1820s, so different even from the earlier days of those skeletal mail coaches, has its brush with fate. As the whirl of wheels, rushing steam, and clanking engines all announce, the railways are coming.

Writing sixteen years before Coketown in *Hard Times* (1854) and experimenting with evocations not yet calcified into Dickensian cliché, Dickens balances Birmingham's industrial progress against its effacing of human agency and identity. The workers vanish into the wheels and noise of steam machinery, an invisible 'hum of labour' running even through the night, work schedules reset to the seemingly indefatigable machines instead of that 'punctual servant of all work, the sun' (1.5). For Pickwick, coaching into town in 1828, this steam-driven industrialization was something only those in the mills were experiencing directly. For the larger audience that constituted Dickens's readership in 1836–7, and especially the middle classes, it was not factory industrialization but passenger railway travel that was threatening their first collective interaction with the new steam-driven machines. And, in distinction from the stationed whirl of Birmingham's manufactory wheels, Pickwick's coach rolling into Birmingham helps to imagine how the Pickwickians rattling along in their coach belong, like the people on the ghosts of the old mails or like those beginning to ride the rails, to a present unified in a historical contemporaneity delimited by the passenger transport system networking them together.

Old Weller embodies this temporally shared world-in-motion the novel choreographed for its first readers. In 1837 *The Quarterly Review* even protested that the type of coachman depicted in Old Weller had already been superseded by the 1820s, and that Dickens, plagiarizing from Washington Irving's *Sketch Book* (1819–20), was not getting it quite right: 'The fact is, the old race of coachmen were going out...and were altogether gone before Mr. Dickens's time. The modern race are more addicted to tea than beer; the cumbrous many-caped great coat is rapidly giving way to the Mackintosh; and, with the change of habits and increase of numbers, they have been doomed to see their authority over stable-boys and their awe-inspiring influence over country people pass away.' In reality, Old Weller is 'a hybrid between two types', as Humphry House carefully discerned.[26] Like the coaches themselves, Mr Weller thus bodies forth the uneven process of technological change, in which the new coexisted with the rapidly outdated, which Dickens saw.

The novel returned to the Old Wellers of the 1820s because, retrospectively, such coachmen were the face of the first generation of public figures defined by the technological acceleration of the transport system. In this, these coachmen were distinct from the more isolated technological generation that had been produced earlier by the forced retirement of cotton spinners and hand-loom weavers. Initially made heroes by the transport network's expanding reach of swift interconnection, the modern stage coachmen then became ordinary by the continuation of that same process (they were 'doomed to see...their awe-inspiring influence over country people pass away'). Then, with the arrival of the railways, they—along with the human face they gave to the connective network—startlingly and rapidly became extinct. Or rather, they momentarily lived on, but only figuratively, as heroic emblems of the historical nature of the present for a community defined by its linking together by an accelerating transport network.

No wonder Pickwick is perpetually being mock-praised as 'Immortal' though he is hilariously unfashionable even in his own time. It is not just that the Pickwickians presented what theorist of historiography Reinhart Koselleck calls 'the contemporaneity of the noncontemporaneous'.[27] Rather they represent the timelessness of a time-bounded crew, and *Pickwick* helped to produce a modern felt truth of temporal collective identity that forever we are those that were alive at this moment.

To redescribe this argument briefly in terms of the novel as a self-consciously evolving genre, *Pickwick* enacted through its temporal structuring around stage-coach journeys a historicizing breakthrough that followed on, and re-adapted, Walter Scott's in *Waverley* (1814). In that novel Scott had famously made past historical change freshly felt as the product of vastly complex competing historical ideological forces, including even the often comically warping effects of myths, antiquarian forgeries, and mad misperceptions of times gone by, and his depiction of those conflicting historical forces unfolding in a past present moment helped turn history into a wellspring of novelistic character, scene, and action. Through *Pickwick*, Dickens rightly joked that historical change had accelerated to such a degree that to write Scott's historical novel he need only to go back a few years. Scott's *'Tis Sixty Years Since* had become, comically but truly, less than a decade, part of an accelerated here and now. The present had become the historical past—posthumous papers—in just nine years, the present was historical, people were more than ever before tightly bound together in time as contemporaries, in both senses of the word.[28] This shared historical present tied to a revolution in the public transport system is both a condition that underlies the production of *The Pickwick Papers* and one that it works through.

Pickwick's structuring stage-coaching journeys were not merely an arbitrary marker selected from among many possible historical markers of the 1820s. The novel's contemporizing effect could not have been reduplicated with other, different coeval historical details, like the factories depicted in Birmingham or the pre-Reform, pocket-borough election of Eatanswill. For even though the passenger transport system is a system akin to many other sorts of systems, and even though innumerable distinct systems—political, religious, commercial, educational, etc.—all organize people's movements in time and space in all kinds of different ways, nonetheless only the public transport system takes as its essential element and aim the collective coordinating of people's individual interconnecting journeys in time and space. This specialized content differentiates the passenger transport system from other systems as their content differentiates them from it.

In this regard, without becoming overly philosophical, one might note that the passenger transport revolution's transformation of human movement in time and space operates directly on a fundamental property of the universe: time and space being themselves founding precepts without which, as Kant probed, existence itself seems fundamentally impossible. (As Kant suggested, one can, for example, contemplate taking away one by one all the many defining properties of a person or an object until one arrives at last at nothing except the space being occupied over time, but to remove that would seem evisceration.) Or, insofar as one instead takes time and space to describe relations and not properties (which complicates but does not refute Kant), it may be useful to think of the passenger transport revolution as partaking of a lengthy and much larger process of reconstructing spacetime in terms of the relativity of motion, which eventually took shape as (Einsteinian) scientific fact.

Whatever its wider underpinnings, the basic point is that accelerating interconnecting trips and their speed define for ordinary people their separating distances from each other in space and time, conditioning both their 'near' and 'far' and their 'here' and 'now', and, for this reason, the revolution in public transport also particularly gets to restructure common time and space. This is why, in the two decades following *Pickwick's* publication, the public transport system helped to coordinate a single shared time for everyone in Britain. First 'railway time' and then Greenwich mean time subsumed all the numerous different local times into one single synchronized time. That standardizing of time across long distances began with John Palmer's 1784 reform of the post. Palmer's reform directed the Royal Mail coaches to carry a clock set to London time, kept by the guard in a locked iron box (sometimes calibrated to adjust for time gained or lost in westward and eastward travel). On the one hand, these clocks made the new mail coaches' correspondingly tight schedules so accurate that some people in the countryside along their routes began using the passing coach instead of the sun as a means to set their local (not London) time. On the other hand, the coach drivers and guards self-awarely running to these clocks represented a first gentle propagation of London time out through the network.

Later this book will explore more fully the standardizing of time in relation to passenger transport. At this point, it suffices to mark its dawning salience. As a revolution in public transport sped getting from one place to another and made that quick connection feel continuous and ordinary and predictable, time contracted for people within its reach such that others previously at a temporal distance no longer seemed so, and the community newly being created became newly temporally bound together internally, in the same way that, viewed externally and historically, they came together as contemporaries. By writing a story about the stage-coaching system currently being displaced by the railways and showing the characters it encompassed unified temporally in their compressed historical contemporaneity, Dickens traced a circle around the first people linked together by a public transport system whose increased circulatory speed had begun reframing once temporally separate people as newly sharing the same time. (And, it may be added, now that's us.)

II. Space

As I have suggested, in *Pickwick* Dickens draws a circle around a community newly becoming linked together temporally by a public transport system. I now want to turn to how *Pickwick*'s structuring as a series of rides around an accelerated public stage-coaching system also reveals the way in which the revolution in public transport transformed space so that it too became shared in a new way. In practice, this separation of time and space is somewhat artificial because of their joint creation by the perception of movement or stasis.[29] An accelerating stage-coaching system abridged distances, making people feel that the same number of separating miles were spatially, as well as temporally, diminished. When what was once six hours away became two, far became closer, and that contraction of distance produced an expanded sense of people's shared space, extending collective proximity.

My concern, however, is with a further story about how the public transport system physically transforms space by making individual movement across it shared. Think here of the difference between a timetable and a route map, and visualize, for a moment, a route map. Just as the revolution in public transport standardized movement in time, requiring people's separate local times to be synchronized to railway time so as to coordinate the 'when' of their newly speeding individual interconnecting journeys and thereby eventually helping to create a common standardized time, so the public transport system similarly standardizes movement in space, requiring people's separate individual journeys to be routed through a network so as to coordinate the 'where' of their speeding individual journeys, making shared the perceived arena of people's circulation. This standardizing of routes is now my focus. My argument is that the public transport revolution's merging of individual journeys across a networked space, as Dickens regarded it in *Pickwick*, helped create the vision of 'ideal relations in community' to which readers and critics, whatever they may think of that vision, have rightly seen this novel aspires.[30]

Figure 4 is a spatial representation—a route map—of *Pickwick*. It charts the five different journeys that Pickwick takes that loosely structure his adventures.

This map is not in itself funny, but it does picture a primary joke of *Pickwick*: that the Pickwickians, motivated by 'an insatiable thirst for Travel' and self-importantly believing themselves to be embarking on intrepid voyages into the unknown, never go further than the easy reach and direct roads of a highly efficient, long-distance, public stage-coaching system. Instead of an arduous odyssey of discovery that explores distant elsewheres, the map graphs a series of regional round trips, whose near-arbitrary sequence, occasional retracings, and repeated starts and returns, all help confirm that Pickwick is speeding around and about an area fully reticulated by a public transport system.

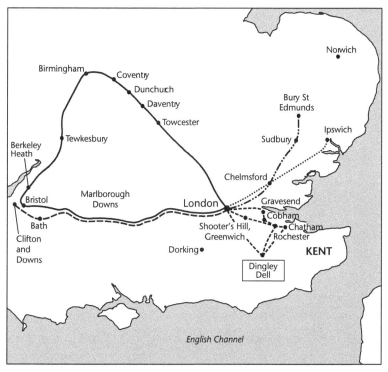

Figure 4. Pickwick's Journeys. Adapted from Robert Patten's 'A Map of Pickwick Tours', Charles Dickens, *The Pickwick Papers*, ed. Mark Wormald (New York: Penguin, 1999), 771.

By definition, a public transport system organizes people's individual journeys in geographic space, merging them by coordinating them through a common network. That is partly what denotes the publicness in a public transport system. This simple fact, however, must have been especially clear, and all the more so because newly historically emerging, when Dickens set out to write his 1820s stage-coaching novel as the railways in the mid 1830s were superseding the long-distance stage-coaching system that had first revolutionized inland public transport. The railways were cutting, embanking, and tunneling an entirely new set of highways, taking 'its iron way—its own—defiant of all paths and roads' (as Dombey scowls), throwing down an entirely new set of stations, and not only combining people's rides on a coach

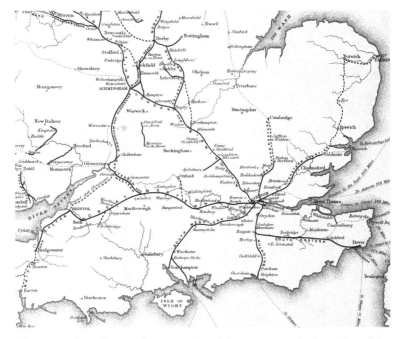

Figure 5. The rail network in 1840. Detail from Francis Whishaw, 'Plan of the Railways of Great Britain and Ireland' in *The Railways of Great Britain and Ireland Practically Described and Illustrated* (London: Simpkin, 1840). Courtesy of The Huntington Library.

but, raising that to another power, also 'training' together the coaches themselves.

In 1836 specifically, with a railway boom in full swing, there was public debate about whether to send the south-eastern railway route from London east through Rochester to Dover, or south and then east (through the Weald of Kent, through Tonbridge and Ashford to the port of Folkestone); parliament was reading a bill authorizing a line from London to Norwich out through the regions of Bury St Edmunds and Ipswich, a 126-mile line that would create the Eastern Counties Railway; the line from London west to Bath and Bristol was under construction, authorized by the Great Western Railway Act of 1835, as was the all-important link north to Birmingham, on the London and

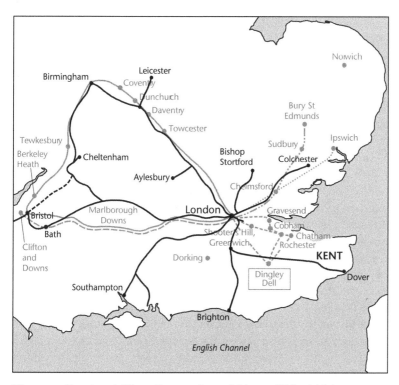

Figure 5. Continued. The rail network overlaid onto Pickwick's journeys.

Birmingham line, authorized in 1833. In other words, unsurprisingly, railways were being planned and built that would soon replace all the stage-coach routes that the Pickwickians travel, except, significantly, their road to the imaginary Dingley Dell, Dickens's mythic Olde England unlocatable somewhere in rural Kent.

Two maps, one showing the rail network a few years after Dickens finished writing *Pickwick* and the other showing that network, minus minor branch lines, crudely overlaid onto the map of Pickwick's journeys, reveal Pickwick's journeys take him almost all around the rail network as it was first coming together (Fig. 5).[31]

Obviously Dickens grasped that the railways were transforming the landscape. Yet, Dickens could also discern—and *Pickwick* shows—how in laying down this new physical network the railways were continuing a revolution in the co-location of journeys into shared space accomplished by the stage coaches in the preceding decades. Dickens thus commonly remarks in *Pickwick* upon the improved roads, the new inns replacing those 'once the head quarters...when coaches performed their journeys in a graver and more solemn manner' (4.90), and so on. The significance of this historic transformation is perhaps hard to comprehend today, attuned as discussions of modernity typically are to the railways' later seemingly more dramatic transformation of space. Taking a close look at one of Phiz's illustrations of this new public coaching system in action can help to clarify (see Fig. 6).

Phiz's picture, like Dickens's novel, depicts a crowded world in swirling motion, and if that was all there were to say, then one might simply nod tritely at how transport has always helped figure the truth that 'life it selfe is but motion' (Hobbes).[32] That bland conclusion would not be wrong or even unimportant. Dickens even counts on it. But it would also not exactly be right. It definitely would not get at what makes Phiz's illustration, and this novel, a picture of the late 1820s as seen from 1836.

In the picture, Pickwick leans out of the speeding post chaise he has hired and chastises Bob Sawyer for his rowdiness. (In a moment Bob's practical joking will remind Pickwick to relax and enjoy the ride.) Sam Weller lunches in the dickey behind, and a postillion steers the horses. In the background a look-alike for Mr Weller drives, whip in hand, a packed stage coach. In the foreground a poor family, labeled Irish by the text and stereotyped as such in the illustration, struggles to keep pace with the rolling chaise, which a strolling traveler salutes on its far side.

At the center is Pickwick's speeding coach. Thinking about the revolution in passenger transport and taking as a given the reciprocal entwinement of humans-and-machines, one might begin analyzing this picture by noticing that most of the passengers' legs on Pickwick's coach, as well as those on the stage coach approaching in the background, disappear here into the transport technology, replaced prosthetically by the near-invisible speeding coach wheels and the horses' legs. The horses' legs highlight the post boy's long left leg, curved to one of the horses powering the chaise, while the plunging hooves, along with the flapping handkerchief-flag and the road-dust clouds, mark the flying speed of Pickwick's chaise. All give meaning to the legs of Bob Sawyer, sitting atop the chaise. Bob's legs, spread

Figure 6. 'Bob Sawyer on the Roof of the Chaise.' Courtesy of The Huntington Library.

'as far asunder as they would conveniently go' (18.533) and blending into the chaise's roof, perform the picture's crowning balancing feat. That balancing feat does not actually testify to Bob's acrobatic prowess, which is dubious at best, since he is totally drunk. Rather, his wondrous, precarious integration into the speeding chaise—he is 'elevated' (18.533), Dickens puns—celebrates, here as elsewhere in *Pickwick*, the glories of the fast-driving 1820s stage-coaching system. The outside passenger, first introduced in the early eighteenth century, represents a kind of human seismograph for the network's smooth functioning: else one would be promptly jolted off. Instead, Bob's easy ride represents a minor miracle deriving from the smoothing power of engineered roads and effective spring-suspension systems, which smoothing also enables the speed exalted here.

Against this picture of swift, successful coaching, Phiz contrasts the comparatively decelerated and struggling pedestrians in the foreground. However—and this is where the picture really begins to reveal its historical complexity—these 'passengers in the street' (2.5), as Dickens calls pedestrians elsewhere in the novel, also form an integrated part of the passenger transport network depicted here. Their naked legs and bare feet do not describe some kind of opposition between nature and technology, not least because they highlight the road as a road. And though Phiz seems to have left (intentionally?) vague whether this road was macadamized and therefore problematically strewn with flint sharp to unshod feet, it is nonetheless clear that the chaotic crowd of people in Phiz's picture, even the baby carried in the backpack blanket-wrap, are, in arrangement with coaches, horses, and roads, all passengers of the passenger transport system—road traffic. And nothing confirms their common identity more clearly—in a tragi-comic way—than the fallen child, at the illustration's bottom, kicking in a futile attempt to keep up with the other foot-passengers. Seen as one element in a comic allegory, the child represents the infancy of human movement in an all-encompassing passenger transport network, with Bob Sawyer apotheosized as coach passenger at the illustration's top.

Consider their coming together on the highway. As part of a public transport system, roads help connect people by standardizing their interconnecting space. They smooth people's common journeys from here to there. The road has thus ever figured as a place of public connection, and it is just barely possible, as the speeding coaches and struggling pedestrians fleetingly negotiate each other, to see the road as providing that connection here. Thankfully, however, this novel does not really believe so much in the road as the crossing place of a world in motion that it expects readers to delight in the sight of a rich man whipping by some poor pedestrians on the trudge. Rather, viewing the road here as a place of interaction tends to dissolve it into the different modes of transport on it, dividing people, and precisely this reminds us that, given solely a network of roads, individuals must (like the pedestrians) still exert themselves independently to get from here to there. People share only the route of passage. Movement across that route remains essentially individual.

What a public transport system offers is to co-locate people's journeys and shoulder that individual exertion. It makes not just the road but *the movement of individuals on the road* shared. And conversely, a public transport system reconfigures physical space to coordinate together individuals' movement—that is, again, it lays down a network that standardizes motion across geographical space in order to allow for merged circulation. We encounter this transformation every time we become passengers moving through a public transport system and, even though we may feel as if we are still pursuing our individual paths, we also find that to use the system we must join our journey to those of the other passengers and conform our journey to the system's routes and stations, to its standardization of our merged movement.[33]

'Cab!' Pickwick cries in the novel's first quoted piece of dialogue, and, as he rolls off on his individual journey via that ultra-modern public conveyance, brand new in the 1820s, the next words out of his mouth are his first destination: 'Golden Cross'—the inn where people meet to catch the stage coach together to Rochester (1.5). In the historical context of the nineteenth-century's revolution in public transport, the speeding coaches in Phiz's picture and in Dickens's novel, backlit by the coming of the railways, no longer represent, invidiously, the different and hierarchical capacities with which individuals traverse the roads. They now stand for just the opposite: an unprecedentedly triumphant combining of individual journeys.

The packed stage coach, in the top left of Phiz's illustration, is thus not just an incidental but a crucial feature. It completes the illustration's layering into three horizontally related planes—pedestrians at the bottom, in the middle the private post chaise that Pickwick has for the first time chartered for this final journey (instead of stage-coaching it), and stage coach at the top. Besides thereby displaying these different modes of passenger transport in relation to each other, this layering showcases the other funny thing about Bob Sawyer, that which at once puts him on a horizontal level with the stage coach and makes him the vertical pinnacle of all three planes. This is that, having joined Pickwick's trip uninvited at its stop in Bristol, Bob is using his legs to take a place on a post chaise as if he were an 'outside' on a stage coach. (In one of Dickens's early sketches, a bride performs something of the opposite stunt when she drapes her wedding shawl over the license number painted on her hackney-coach door to disguise it as a private carriage.)

As an outside, Bob Sawyer along with the stage coach helps ensure that Pickwick's hired post chaise is made continuous with the public transport system—as it certainly was. (Similarly, the stage coach seems to be carrying another Pickwick in the corpulent bespectacled gentleman behind the driver.) Absent Bob or that stage coach, Phiz and Dickens risk readers relating Pickwick's chaise—as it is also related, notice the servant's dickey—too heavily back to the kind of aristocratic private coaching of the sort belonging, as Dickens repeatedly smiles with literary nostalgia, to those old romantic days when ne'er-do-wells absconded with damsels pursued full tilt. In *Pickwick*, when, for instance, an old squire-type like Wardle madly races after a villain mercenarily eloping with his spinster sister, it's an old scene forwarded comically into a new world. Poor rich Wardle will inexplicably find his hired chaise held up at the turnpikes, his fresh horses taking forever to be hitched on at the stage—the whole infrastructure of a public transport system, upon which he actually relies but fails to see, cannily bribed into delaying him by Jingle, whose villainy takes the form of an individual violating the public transport system as a public system.

It is impossible to overstate the importance of the fact that Pickwick and Sam Weller, Dickens's latter-day Don Quixote and Sancho Panza, do not sally forth upon their own steeds. The Pickwickians are all quintessentially passengers. In the novel's second number, Dickens accords the point its own chapter: 'A short one— showing...how Mr. Pickwick undertook to drive, and Mr. Winkle to ride; and how they both did it'—which is to say, how they didn't, couldn't, do either. Instead Dickens treats his readers to a stupendous display of technological incompetence that leaves no mistake undone and no one uninjured. In one classic screw-up, tantamount to popping the clutch with the gas pedal floored, Pickwick tries to calm Winkle's nervous horse by approaching him with a tremendous chaise whip in hand. As these full-time transport-system passengers discover, attempting to assume individual control of the technologies undergirding their coaching adventures would be nothing short of a nightmare: 'It's like a dream...a hideous dream. The idea of a man's walking about, all day, with a dreadful horse that he can't get rid of!' (2.49).

To return, then, to Phiz's picture: What's being celebrated? What is everyone in the picture looking at? The answer, which clarifies this novel as a whole and what it shows about this moment in the history of passenger transport, is a public stage-coaching system that has in the previous few decades freed and expanded and extended and accelerated individual interconnecting journeys, transforming the roads by successfully merging together on them people's individual movement as they travel along their separate paths.

In *The English Mail Coach, or The Glory of Motion* (1849), Thomas De Quincey remembers his ride on a speeding mail coach that brought news of the nation's military victory to town after town and that in so doing constructed citizens and nation en route. Experiencing the linking of the towns and people together even as individuals' reactions to the news varied, De Quincey concludes by imagining the nation springing into being through the network of moving stage coaches in which his journey has traced just one trajectory. By contrast Dickens's novel takes the stage coaches not as intermediaries for politics or nation but as unifying in their own movement. Bob Sawyer has made a flag of his silk handkerchief for no other purpose than to celebrate their stage-coaching journey. *Pickwick* makes that flag wave.

One can also now begin to see why it matters that Phiz's picture captures a moment en route between towns (from London to Bristol), why the reach of Pickwick's journeys on the map is regional, all intercity routes—why, in short, Dickens needed to send Pickwick out of London. Besides thereby keeping Pickwick continuously on the go and making the coaching inns, featured by the dozens, a primary setting of the novel, the stage-coaching system's interlinking of separate cities with a continuous, fast, and convenient public-transit system revealed that ordinary people, who had always traveled between those cities, had become, for the first time, no longer exclusively bound together and separated spatially by the static geographic places in which they lived. It was not that the people who lived in Bristol now lived in London, though there are plenty of accounts from the period, especially after the railways, about how the acceleration meant everyone suddenly felt next door to each other—'as distances [are] thus annihilated, the surface of our country would, as it were, shrivel in size until it became not much bigger than one immense city'.[34]

Rather, what the stage coaches, and then the railways, made clear was that the fast network connecting those two separate places now mattered separately from geographic distance. Where one lived still counted, but it now counted differently. Alongside the collectives defined by their actual physical proximity, the public transport network also defined one based in a shared circulating movement. We all now live in this warped space. Today the major cities with their fast air links or high-speed rail may be more closely and conveniently connected to each other across hundreds of miles than to towns lying miles nearer geographically. So while London would experience its own internal public transport revolution—the omnibus in 1829, underground metropolitan rail in 1862—the intercity routes Pickwick rides established the public transport system's swift interconnections as a means to defining a new form of community based on people's networked arena of coordinated circulation.

So, taking one last view of Pickwick's coach in Phiz's illustration, imagine Dickens, writing with the new railed roads' tracks humming with approaching steam-driven iron wheels, focusing his readers on how individual journeys upon the road had previously successfully been coordinated in space into a public transport network. Compositionally, the half-hidden rolling coach wheel in Phiz's picture, which competes for center place with Pickwick, internalizes, within the picture, the picture itself as an unframed circle. That coach wheel might thus be described as carrying the whole rolling world of Pickwick's adventures, and how that could be, how this notoriously formless story thus could be understood to be structured by its coaching trips, rather than excused by them from anything resembling structure, is the next question addressed here.

As every minimally competent passenger of public transport knows, in making movement on the road shared, in merging individuals' journeys, the public transport system does not turn everyone's individual journey into a common single journey. Rather, it routes individual journeys through a standardized space, into a delimited network, where people get on and off at different stops, use different lines, and so on. The crucial, related point that needs now to be taken into account is that this standardization of space is thus premised on serving a bustling, circulating society in which all kinds of individuals are making journeys for all kinds of individual purposes to many different places, and the people entering this network are all filled with their separate, different purposes. While these passengers—many are journeying on business, as commercial travelers—certainly may relate to their journey's individual purpose in all kinds of different ways, including seeing it as their own or repudiating it, they are not traveling to travel. They are using the network. The system provides an infrastructure that helps them carry out their purposes, much as they ordinarily use the communication system to send their messages.

The Pickwickians thus ride the public transport system differently from most other people. Instead of entering the transit system as part of making some individual journey for some individual purpose, the Pickwickians ride the stage coaches simply to travel around and about. One can see this, for example, in the next scene illustrated by Phiz after 'Bob Sawyer on the Roof of the Chaise', in the stop the Pickwickians make at Towcester on the road back to London from Birmingham, when they run into some ordinary travelers. At a coaching inn, they bump into their old friend Pott, the newspaper editor from Eatanswill. Pott—whose politics are staunch Blue—is traveling on a mission to crash a Buff political ball being held in Birmingham. 'WE WILL BE THERE', he has editorialized (18.548). After a happy reunion with the Pickwickians, Pott discovers his arch-rival editor Slurk is staying at the Inn, also en route to the ball. In an uproarious scene the two finally go at it. What matters here though is how even these staunch enemies at complete cross-purposes, who in fact are utterly defined by their always being at cross-purposes, so naturally resemble each other in using the stage-coaching system to get somewhere for their separate individual reasons. By contrast Pickwick, who for the first time actually had something like a reason for making this trip, immediately turns Winkle's request to deliver a letter to his father in Birmingham into an excuse to enjoy one last road adventure.

There is no particular adventure the Pickwickians set out to have, and Pickwick's many adventures are all ultimately subordinated to the purpose of traveling around to have them. Whenever some special purport, some specific direction, some plot, threatens to arise in the course of those adventures and overtake Pickwick, Dickens again and again melts it away into the larger, seemingly aimless aim of stage-coaching around, often simply by keeping on going on. This is quite obvious, and nowhere perhaps is it made more so than at the moment when the plot seems thickest, and, having been found guilty of breach of promise and accorded two months to pay his fine, Pickwick rejects paying (thus ensuring his future imprisonment), further rejects any discussion of his predicament, and instead smilingly announces to his friends: 'And now . . . the only question is, Where shall we go to next?' (13.371). *Where shall we go to next* is, it has been well understood, the essential logical structure of this novel, but one easily and often mistaken for not much structure at all, or pre-empted by the genre label 'picaresque' into nothing more than sequential road encounters. Yet this logic, one can readily see, frames the story as organized by its five stage-coaching journeys. This is why when one looks at the map of Pickwick's journeys one does not merely see the route a character traced as his travels unfolded, but rather a view of the structuring of the novel, confirming that whatever adventures he had along these routes, Pickwick's traveling around this stage-coaching network has been his purpose.

Again, this is odd: because the public transport system works by subordinating itself to everyone's individual purposes, offering to shuttle people around as they go about their business, Pickwick appears purposeless to others, and sometimes to readers. Similarly *Pickwick* itself seems aimless. This formlessness, however, actually has a specific meaning; it follows from a shaping historical perspective; and what Dickens is continually showing in this novel, as well as the way that novel presents Pickwick as a character, is historically contingent upon imagining a hero emerging from, and in a sense standing for, a revolutionarily public transport network. Because people understand this system to have arisen to help them accomplish their individual purposes, a protagonist (like Pickwick) fully identified with this system would necessarily appear (like that system) both purposeless and perpetually in danger of being swept into the various purposes, the more various the better, of the people he arbitrarily encounters as he rides the system.

In thus fully identifying Pickwick with the structure of the new public transport network, which after all has endowed him with his name, one can suddenly better comprehend the larger structuring pattern of this novel. Perhaps most sharply, one can see why and how it makes sense that Pickwick, ridiculously, ends up on trial, drawn into a plot that completely misrecognizes his intentions and actions, making up for him purposes he never had. The trial, that central misrecognition of Pickwick's intentions, like the endless confusion the Pickwickians provoke throughout the novel, is not a sign of wild creative formlessness, Dickens's creative genius set loose. Rather, it shows Dickens brilliantly sharpening the perplexities of taking as a hero a passenger-avatar of the public transport network. Only readers get to see the gap between *Pickwick* the novel, in which Pickwick purposes to ride around the regional passenger transport system, and the diegetic story world in which he repeatedly encounters people unable to grasp his purpose because that regional passenger transport system is defined, infrastructurally, by its subordination to people's various individual purposes. One can thus read this story as a series of hilarious collisions between this hero who necessarily appears purposeless, though he is not, and the people, full of their different purposes, that he encounters adventuring around the public stage-coach system of which he is the tutelary genius.

In this structure, stock literary scenes of any kind become fodder for Dickens's story. The whole point is that Pickwick shrugs off established modes of characterization because they carry the promise of purposefulness, and that at most he gets awkwardly and temporarily and comically enmeshed in stories filled with recognizable purposes that he does not fit. Traditional romance plots especially threaten to rush in where there would seem to be no other meaning. The novel as a whole partly garners its shape by corralling its romances into subplots that paradoxically matter partly because neither the narrative nor readers are supposed to care much about them. Arabella and Winkle? Snodgrass and Emily Wardle? The street directions to his house that Bob Sawyer gives in loving detail to Pickwick preoccupy this novel a thousand times more than the opening of these romances, which register instead by the way in which they affect the men's behavior on their coach ride (and never mind the women's perspective altogether). Meantime—and it almost feels too obvious to say—again and again Dickens will delight in creating circumstances or imagining characters that project romantic or even salacious purposes onto this innocent protagonist of the public transport system, so that as Pickwick travels this system purely for the sake of traveling around, he will comically engender a kind of shadow Don Juan Pickwick: seducing his landlady, rakishly appearing behind the door at a girls' boarding school, a libertine snuck into a lady's bedroom at an Inn: 'Wretch!' (8.234).

This last scene offers a wonderful extended joke in which Pickwick wanders around a coaching inn unable to figure out which bedroom is his. And Pickwick's mistaking of another person's room for his own nicely foregrounds the structuring logic of the public transport system by which the people journeying in it are fully engaged with their own individual purposes, while at the same time the system speeds them along their journeys by standardizing the space through which they travel, making their movement interchangeable, available to any individual, creating, for pointed instance, rafts of identical rooms for people to stop over in.

Here is the narrator describing Pickwick, who has forgotten his watch downstairs, leaving his bedroom to retrieve it: 'The more stairs Mr. Pickwick went down, the more stairs there seemed to be to descend.... Passage after passage did he explore; room after room ... [at last] Mr. Pickwick seized the watch in triumph, and proceeded to re-trace his steps to his bed-chamber.... [But] his journey back, was infinitely more perplexing. Rows of doors, garnished with boots of every shape, make, and size, branched off in every possible direction. A dozen times did he softly turn the handle of some bed-room door, which resembled his own, when a gruff cry from within of "Who the devil's that?" ... caused him to steal away.... He was reduced to the verge of despair, when an open door attracted his attention. He peeped in—right at last. There were the two beds, whose situation he perfectly remembered...' (8.232).

As he undresses and slips into his bed, Pickwick chuckles over his confusion, 'Droll, droll, very droll' (8.233)—and what is slightly amusing is the surrealness of the place, which in essence dramatizes the standardization of space by a public transport system. Individuals are, from the system's point of view, necessarily interchangeable—all the bedrooms look alike; who can tell which one is even mine? All 'droll, very droll', but then: 'Mr. Pickwick almost fainted with horror and dismay. Standing before the dressing glass, was a middle-aged lady in yellow curl-papers, busily engaged in brushing [her hair]' (8.233). Where before readers observe along with Pickwick the way in which a coaching inn standardized space into a labyrinth of identical rooms, now Pickwick turns out to have, as it were, crossed right into the dimension of space he and the reader imagined he was observing.

Pickwick's shock here might be thought to stem from discovering that his interchangeability with other individuals extends right into that most personal of private spaces, the place where one wears a nightcap—even those beds whose situation he perfectly remembered. But what the scene really brings home is that the shocked horror Pickwick feels arises from the surprising fact that even though the system lays out the space of the stage-coaching inn to be used inter-

changeably by different individuals, such that one might understandably end up in the wrong bedroom, nonetheless the individuals moving
through that space do not actually become undifferentiated. On the
contrary, they are all moving through that space on their individual
paths, with their individual purposes. And so, Pickwick finds himself
not only accidentally to have invaded someone else's room, but also,
much more fearfully and overwhelmingly, to be entrapped in another
narrative, thoroughly structured by gender and genre, a lady, undress-
ing, a bedroom, night: 'Wretch! . . . what do you want here?' (8.234)

One reason, then, to believe that Dickens gets this scene just right is
that, having set up the reader to see the interchangeability of the individual travelers from the system's perspective, he then allows the reader
to experience with Pickwick the jolts of perception by which he first
thinks *his* room has been invaded, then grasps that he has invaded
another person's room, and finally recognizes that, far from occupying
these individually interchangeable spaces interchangeably, he has,
through his mistake, become enrolled in a completely different story,
with completely other purposes than his own.

From here, this mishap simply dominoes. One need not believe for
a second that when Sam rescues Pickwick from the hall outside the
lady's room Sam truly suspects Pickwick of the rakishness Sam delights
in implying. But it feels comically inevitable that the next morning
Pickwick will continue his counseling of Peter Magnus, a man he has
met on the coach ride up, about how to propose marriage to the
woman he has traveled to meet, and then that, after Peter proposes,
the lady in question will turn out to be the middle-aged lady in yellow curl-papers. Poor, annoying, quickly forgotten Peter Magnus!
Brimming with purpose to propose to the lady in yellow curl-papers
(Witherfield is her name), he has randomly crossed paths with Pickwick on the coach ride to Ipswich. In another Dickens novel, that
accidental meeting of two fates might reveal hidden social relationships connecting their seemingly separate fates. This hero descends
upon Peter Magnus out of nowhere for who knows why. (Who sees
the public stage coach as a place from which to come? It is for getting
places.) All Peter Magnus can see is that Pickwick and his beloved
have met, but they won't confess to how. There must be a dark plot,
he feels. But he can find no plot. Just as his lady had cried 'Wretch!
What do you want?' and Pickwick responded, 'Nothing, Ma'am—
nothing whatever, Ma'am', so poor Peter Magnus will threaten 'you

shall answer it, Sir', only to have Pickwick demur, 'Answer what?' (9.246). Pickwick is not giving him the runaround. He really is just running around.

When Sydney Carton, however purposeless had been his life, rides a cart to the guillotine at the end of *A Tale of Two Cities*, he truly has interchanged his fate with Charles Darnay, who is unconscious in that coach fleeing Paris. Pickwick, by contrast, takes his meaning partly from seeming to have no fate to interchange. He is, if anything, the kindred spirit of the coaches rolling people to their fates. Thus the conflict between Peter Magnus and Pickwick only gives rise to a sham plot, not a real conflict of purposes, and Pickwick will ludicrously be hauled in front of a magistrate for threatening to fight a duel over a lady. This is this novel's essential structure at work: the hero falls into a mistake related to his coaching journey, but partly because of the invisibility of that journey as a purpose, he suddenly finds himself insanely embroiled in some other person's plot.

As should also be clear, there is no way to resolve this conflict, since the point here is that it is not actually a conflict of purposes. Following the story a bit further to notice how Dickens extricates Pickwick from his predicament confirms as much. What happens next is that Jingle turns out to be imposing on the magistrate Nupkins's daughter and family, pretending to be a captain, and Pickwick exposes him. It might seem as a result that the episode with the lady in yellow curl-papers gets dissolved by redirecting attention to another plot. And like another poor Peter Magnus, one might mistakenly conclude one has found Pickwick out: Here! There was a plot brewing! Pickwick has triumphed over his enemy Jingle!

But who exactly is this enemy Jingle? What is this plot? Dodson and Fogg, we understand: they are Pickwick's enemies because they intentionally ascribe to him a criminal purpose where he had none, making him look as if he has been plotting when his life is all anecdotal, and they bring him into an irresolvable moral conflict. Jingle is different. Jingle does not plot against Pickwick, and yet only he is sporadically pursued by Pickwick as Pickwick otherwise aimlessly travels around the public transport system. He materializes out of nowhere when Pickwick catches his first stage coach; his intentions seem hard to discern; he too acquires a sidekick—Job Trotter—mirroring Sam Weller, and he similarly circulates throughout the book, also apparently aimlessly, around and about the same terrain.

Jingle is, in short, Pickwick's doppelgänger, and what specifically that means is that instead of appearing oddly purposeless to the people Jingle meets as Pickwick does, Jingle—inverting Pickwick—parasitically shapes himself to whomever he meets and nefariously infiltrates himself into their lives. To poet Snodgrass's inquiry whether he is a 'Poet, Sir?' Jingle will lie: 'Epic poem,—ten thousand lines . . . bang the field-piece, twang the lyre'; to romantic Tupman's demand, 'Many conquests, Sir?': 'Conquests! Thousands. . . . Donna Christina—splendid creature . . . prussic acid—. . . romantic story—very'; and so on (1.9, 10). As Pickwick's doppelgänger, Jingle resembles so as to reverse, and what, among other things, Jingle is running backward is, amusingly, Pickwick's purposelessness. Hence without having any really clearly discernible aims of his own (it's important that not everything he does is for money), this strolling actor will continually be discovered around and about intently misleading other people about his purposes. In particular, like an evil shadow of Pickwick's innocent shadow, he will repeatedly actively fake performing the character of lover, which character Pickwick repeatedly finds it so difficult to repudiate being perceived to be truly performing. To others Jingle appears simply an imposter, but to Pickwick he is an imposter of Pickwick's own unwitting impostures. Pickwick says to magistrate Nupkins: 'I have every reason to believe, Sir, that you are harbouring in your house, a gross impostor!' 'Wretch', cries Mrs Nupkins to Jingle—and one hears the echo (9.260, 267).

Understanding this doubling relation, one can see why Pickwick's running into Jingle at Nupkins's does not really present the reader with a plot in the ordinary sense, though the pursuit of Jingle has brought Pickwick to Ipswich. The entire sequence looks more like an impromptu chase by the mythic hero of the public transport system after his nemesis undertaken through that system. Pickwick bumps into Jingle at a literary fête in Eatanswill; with hardly a goodbye (since it's not as if the content of any adventure matters as much as the going), he pursues Jingle back to Bury St Edmunds—'I will follow him!' (6.157)—gets tricked there by him and loses him. He returns to London, almost immediately learns from Mr Weller (who as a coach driver naturally knows where people in the system are) that Jingle is in Ipswich, and—again, spontaneously, 'I'll follow him!' (7.206)—Pickwick is off to Ipswich. All by stage coach. No matter that once upon a time not so long ago places like Ipswich and Bury St Edmunds, which were always relatively near to each other, were both relatively far from London. It's all one big transit network now, and Pickwick is its 'reg'lar thoroughbred angel', as Sam calls him, verbal horseplay intended (16.488). On the Ipswich coach Pickwick encounters Peter Magnus traveling to propose to a lady, accidentally ends up in the wrong bedroom of that lady, is taken (rakish rogue that he seems to be) to a magistrate's where—the reader by now having completely forgotten why Pickwick originally came to Ipswich—surprise, he rediscovers Jingle, the evil shadow he had been pursuing, now impersonating a military captain in order to seduce Miss Nupkins. 'I have every reason to believe, Sir, that you are harbouring in your house a gross impostor!' 'Wretch!' Echoes. And now the doppelgänger Jingle has his say: 'Ha! ha!...good fellow, Pickwick—fine heart...see you again some day— keep up your spirits—now Job [Trotter]—trot!' (verbal horseplay intended) (9.268).

Here, as throughout the book, the shaping structure, the seed, of all this wonderful escalating, apparently incoherent inanity is that Pickwick bodies forth a fundamental disjunction, or paradox, of the public transport system exemplified in those interchangeable bedrooms in the coaching inn: these foreground, as Pickwick—both the book and the character—foregrounds, how the rise of a public transport system transformed people's journeying, making it shared and thus making all individually interchangeable in its system, while at the same time the system was subordinated to its individual users' purposes, ensuring that its riders were not at all interchangeable, but on the contrary were individuals each pursuing their individual ways.

Recognizing this structure helps to provide a historical account of why *Pickwick* has struck so many critics and readers over the years, whether they have deemed it successful or not, as being about the generation of 'ideal relations in community'. It also explains why, as John Bowen says, so many have been torn between recognizing that the materials of Dickens's story feel completely accidental and contingent, and yet that the story itself somehow nonetheless feels transcendent, unified— what could possibly gather together such diversity? It's simply, as Chesterton quotably gushed, 'the gods gone wandering in England'.[35]

The reason *Pickwick* has been read this way and even created this feeling in those many readers is that it does not found its community on any particular identifying content or shared purpose, but upon the rise of a public transport system that triumphed by embracing people's movement on the road together in a network, making their journeys shared in space and time, even if their purposes were as different as a Pott's from a Slurk's.[36]

This is not to suggest that Dickens represents the public transport network as some kind of transcendent system radiating uplifting harmony. Even in *Pickwick*—especially in *Pickwick*—the public transport system's operations offer virtually endless resources for conflict and pain. Immediately upon departing from Golden Cross, Jingle confronts Pickwick with the anecdote of the mother who forgot to duck under the arch: 'Crash—knock—children look round—mother's head off—sandwich in her hand—no mouth to put it in—head of a family off—shocking, shocking' (1.9). Nor is David Parker wrong to read the novel's extensive dealings with horses as reflecting a divisive class terrain that, as he argues, works 'to mark Mr Pickwick's position in society, to indicate differences in customs and experience between different social groups, [and] to reveal hurdles social ambition needed to jump'—all at a moment when the railways were demolishing equestrianism as an ancient class prepotency.[37] Nonetheless, much more definitively, *Pickwick* engages the public transport system's untranscendentally material role in linking together people's circulation in space and time.

Shifting the grounds for community away from content itself, this novel strives toward a social cohesion that overcomes the divisive fissures inevitably produced by individuals or collectives oriented to, or defined by, any certain purpose. Over and over again with comic inevitability the Pickwickians become embroiled in turmoil and hostilities, the repeated overcoming of which produces the novel's triumphal comedic viewpoint, and yet, as Hans Robert Jauss sharply observed in his Freudian argument that *Pickwick* exalts the ego's capacity to dismiss life's horrors through humor, the Pickwickians never overcome conflict by resolving it in terms of its subject matter.[38] Instead, Dickens produces one recalcitrant contention after another so as to introduce repeatedly the solvent of a public transport system—Pickwick—that ignores potentially curdling content as merely contingent.

No wonder why actually reinscribing the Pickwickians solely in terms of the specificity of their identity—impounding them as the all-male, white, decently well-off London tradesmen that they are—instantly destroys their power to generate universal comity. This novel's vision of community is not trying to erase, or resolve, existing social divisions or such identifications. These remain everywhere: in the ever-present class distinction between Master Pickwick and his manservant Sam; in the unsympathetic marginalization of a single mother like Mrs Bardell; even in the passing synecdoche for colonial oppression of that poor Irish family on the trudge, whose father surely might have been found waving his walking stick quite differently elsewhere, and definitely not cheerfully or in synch with a better-off stroller across the way. Such refocusings are not wrong. Neither are their disenchanting effects unreal. But they are not Dickens's superintending viewpoint in this novel where the stage-coaching system links individuals on the go such that even the real fences of an actual pound ultimately cannot divide the newly networked community: 'How long this scene [of Pickwick being pelted and jeered in the wheelbarrow in the pound] might have lasted, or how much Mr. Pickwick might have suffered, no one can tell, had not a carriage which was driving swiftly by, suddenly pulled up...' (7.197).

As that early scene of Pickwick in the pound reminds, this novel later similarly confirms the stage-coaching system's power to undergird community with a special trip by Pickwick that does not appear on his route map. This 'tour', the only one that Dickens actually calls a 'tour' instead of a 'journey', is extremely short and not done by coach, and it occurs in the chapter titled 'Mr. Pickwick makes a Tour of the diminutive World he inhabits, and resolves to mix with it in future as little as possible'. In that chapter, Dickens briefly recounts Pickwick's last look around that systematic impounding of humans, the Fleet prison: 'From this spot Mr. Pickwick wandered along all the galleries.... The whole place seemed restless and troubled; and the people were crowding and flitting to and fro, like the shadows in an uneasy dream. "I have seen enough" said Mr. Pickwick... "Henceforth I will be a prisoner in my own room"' (16.490).

In the terms explored so far, one can readily understand the important oppositional role of the Fleet prison episode within the novel. It is a public system for standardizing space to halt the circulation of individuals in time. Upon first entering the Fleet prison, Sam notices a Dutch clock and then a bird-cage, and, with memorable concision, he registers these two objects' emblematic relation to their surroundings: 'Veels vithin veels, a prison in a prison. Ain't it, Sir' (14.434). Sam, as usual, has it exactly right. In the Fleet, time distorts into its passing on the wheels of a clock as human life runs out (literally, there being no limit to the time served by debtors), while space collapses into a cage barring life's course. The Fleet prison is the antithesis of the idea of community celebrated in *Pickwick* because it represents the paradoxical fact that community is here being defined against the spatial physical clustering of everyone together as a collective, each with their assigned place, their chummage ticket, so that, conversely, community, as configured by the passenger transport revolution, means interconnected individuals 'going' whithersoever they choose along the network unifying them. This is why the Fleet's horrors are both too much for Pickwick, who retreats into his cell like a useless coach into its shed, and also the novel's central space of negation that allows Pickwick to overcome his deepest conflicts, with Mrs Bardell, with Jingle, not by resolving those conflicts but simply by setting all free—that is, in this novel, back into the network of individuals circulating.

To discover beyond this what the Fleet episode reveals, one must look hard at the fact that Dickens structures Pickwick's encounter with the prison system in symmetrical opposition to, and inversion of, the public transport system as he has been depicting it (and personifying it in Pickwick). Most importantly, where before, unlike everybody else, Pickwick rode the public transit system with no purpose at all, personifying its mission of linking together whatever is going on, now Pickwick somersaults. He is suddenly filled with purpose (of standing up to Dodson and Fogg's chicanery), and he makes his imprisonment his choice and intentional act—in this, again appearing unlike everybody else in the system that he is in.

Much that occurs in the Fleet episode follows. To engineer the six chapters that stretch from Pickwick's arrest up to the arrival of Mrs Bardell and Pickwick's exit from the prison (i.e. from 'Introducing Mr. Pickwick to a new...Drama of life' through his final 'Tour of the diminutive World'), Dickens interweaves three narrative threads. In one extended parody, Sam exposes the comic dimensions of Pickwick voluntarily entering a system whose purpose is to block the carrying-out of its inmates' purposes. First, Sam lays a secret plan to make his father his creditor and get himself arrested. Then, in a culminating moment, Sam relays to Pickwick the funny and brilliant parable about holding to purposes and principles in which a crumpet-eating man shoots himself to prove that he (contra his doctor) was right that eating three-shillings-worth of crumpets wouldn't kill him. Finally, in a concluding scene, Sam remains comically unrepentant through a set-piece in which pastor Stiggins and Mrs Weller bewail his guilty fall. In another narrative thread, Pickwick discovers Jingle. While the two ostensibly remain inversions of each other in that Pickwick is now filled with serious purpose and Jingle is seriously ready to give up the ghost ('Nothing soon—lie in bed—starve—die', 'all over—drop the curtain', 15.454), in fact the Fleet prison's inversion of the passenger transport system from which Pickwick and Jingle took their meaning puts an end to their opposition. 'Take that, Sir', Pickwick says, giving him money (15.454). The Fleet, depicted as the blight of all activity and action, fails to provide a field in which Jingle and Job can even pretend to act. Having sold their wardrobe to eat, they diminish to their diminishing bodies; the curtain is dropping—'There is no deception now. . . . *these* sort of things are not so easily counterfeited . . . and it is a more painful process to get them up', says Job, indicating his sunken cheeks and his bone-thin arm to Sam (16.487).

This assertion by Job that the Fleet prison represents hard reality, putting an end to all previous make believing, points to the central thread of the Fleet episode: the series of descriptions through which Dickens the reporter uses Pickwick as a means to expose how the Fleet prison system works and how it inhumanely crushes its inmates— not just back in 1828, but there and then in 1837. In scenes calculated to lay before readers the hidden horrors of the Fleet prison, Pickwick first encounters the prison's pain-filled galleries, faces his own loss of privacy, dignity, and independence in a hazing from his raucous drunk roommates, and, having learned the rules of chummage, rents a private room ascribed by seniority to an utterly miserable, dying Chancery prisoner. After this initiation and wised to the fact 'that money was, in the Fleet, just what money was out of it' (15.450), Dickens conducts Pickwick (with readers in tow) into the animate charnel house, the 'poor-side', where the inmates without any money starve. This narrative thread divulging the Fleet's horrors then culminates in the ultimate 'discharge' of the Chancery prisoner, one of Dickens's most unaffected deathbed scenes. Shortly thereafter an anguished evocation of the prisoner's dead body, 'all quiet and ghastly' lying in a shed (16.489), followed by a bleak short shot of hard alcohol, brings Pickwick to take one last brief look around and then lock himself in his room—'I have seen enough . . .'.

Enter, it would seem, reality. 'This is no fiction', Dickens declares outright at one moment. All the suffering seems simply all too real. While every novel reader knows that to prop up the suspension of our disbelief one of fiction's most common fibs is 'This is no fiction', in these scenes in which Dickens censures the workings of the Fleet the seams of the comic fictional world really are coming apart.

As readers often remark, the Fleet episode blatantly seems to disrupt the coaching story as a story. Put most simply, the considerable difficulty posed by the journalistic Fleet episode is that insofar as Dickens presents the Fleet as an encounter with reality, with hard truths, then the stage-coaching story suddenly seems to stand by contrast as nothing but an escapist fantasy, something merely made up by Dickens. And perhaps at one level Dickens-the-crusading-reporter, worked to the boiling-over point by flagrant injustice, grants that dismissal. But at a deeper level, Dickens the novelist certainly does not mean by asserting the realist cliché that 'This is no fiction', or detailing some of the harrowing conditions in the Fleet, or discombobulating his fiction with some reportorial prose, that one should dismiss all that has come before as solely a contrived fabrication. On the contrary, the glories of stage coaching are no less real to Dickens than the Fleet prison, which in fact he had never visited.[39]

Rather, Dickens is engaging, as he does within all his novels, in examining the real power of fiction, and specifically here he is confronting his having rendered a public transport system that is meaningful partly because it is indifferent to his characters' purposes and unconcerned with what particular plots unfold within its aegis. That overall scenario—in which an author controls character and plot not in order to relay through them his story, but to show off a public transport system that underprops almost whatever diverse characters and plots he introduces—might be said to be paradoxical or contradictory in purposing to show the system's purposelessness, but it has not been relevant to entering into, or understanding, Dickens's fictional stage-coaching story. Up until the reader reaches Pickwick's incarceration in the prison system, the reader has not had to worry about this odd fact that lurking behind the stage-coach adventures is a controlling author.

The Fleet episode changes that. The Fleet prison and Pickwick's imprisonment there, interrupting and inverting the coaching story, introduce into the novel the depiction of characters caught in a system defined by a controlling authority that is intentionally indifferent to the individual purposes and plots of those within it. The Fleet episode thus enfolds within the novel the contradiction, which Dickens knew intimately but that readers had merrily been able to ignore, that a controlling author has been intending to demonstrate a passenger transport system that does not care about his characters' purposes or plots. Now a prison system mirrors, in inverted and dark colors, the hitherto suppressed reality that the characters are not after all quite the adventurous independent agents that readers imagined and also that the coaching story—the novel's structuring in terms of an infrastructure of transport—that the reader had accepted as holding together all their various adventures is not quite the stand-alone structure that it seemed to present itself as.

Instead, enfolding the logic of his narrative's form into his story's contents, Dickens registers not that the stage-coaching novel is 'made up' in the sense of untrue, but that, in being 'made up' in a novel, the characters, even those who never meet, garner a certain invisible, bounded connectedness from the fact of their arising out of an intended act of authorial imagination. No matter that their creation is spread out over time, or that their story is chopped into separate physical parts, or that they are jointly constituted by an illustrator, an editor, and so on, insofar as readers take them to be in a novel together these characters are connected. Phiz caught the significance of this unity when after the serial run he had to give the story a frontispiece and pictured a scene—not chronicled in the story—of Pickwick and Sam at a table together enjoying a book that presumably recounts their own adventures. Thus does it all hold together.

What the Fleet episode shows, then, is that this novel is not merely an absorbing fiction that reflects how a revolution in public transport engendered a new kind of community. Now the secret and real contribution of this novel as a novel comes out: the reader's experience of *Pickwick*'s characters as sharing a common arena of circulation that flows from the imaginary act of Dickens's authorship helps create here the experience of shared circulation in the public transport system.

Picture yourself on a speeding coach, a railway car, in a public transit system. You are always local, always moving or stopped, and generally you are intent on getting somewhere to do something. Picture again in your mind Phiz's picture of Pickwick's coach with Bob Sawyer on top—always local too. Neither the characters inside *Pickwick*, including Pickwick who personifies the system, nor the readers outside the novel necessarily experience their mobility as a shared temporal and spatial arena of circulation simply by looking at the person next to them, or by riding around the system, or even by looking at a timetable or transit map. On the contrary, wouldn't witnessing—especially at first, in the early nineteenth century—the rapid and wide trajecting of people here and there have been more likely to look like the tearing up of community? One would have to imagine the real relations one had to others in this network by virtue of its being a network. And in the nineteenth century the chief place where one could experience such imagined synthesis of networked relations was in the novel. This was *Pickwick*'s contribution.

This is not to claim that if *Pickwick* had never been written people would somehow never have recognized that a revolution in public transport had bridged separation over distances that previously defined different, separate communities to create instead an imagined community undergirded by its network. But it is to push beyond *Pickwick*'s 'reflecting' or 'figuring'—or anything of that adjunctive mimetic order—the public transport revolution. Through a technology of the novel, *Pickwick* helped instate the experiencing of the public transport system as creating a new kind of community in people's circulation around a network. This was a part of the historical, ideological work it accomplished, as well as a key to its successful narrative form.[40]

It is thus perhaps worth recalling at this point that *Pickwick* was received in its own time and place as a cultural phenomenon. Long before its serial run ended, reviewers were agog: 'in less than six months from the appearance of the first number of the Pickwick Papers, the whole reading public were talking about them—the names of Winkle, Wardell, Weller, Snodgrass, Dodson and Fogg, had become familiar in our mouths as household terms. ... Pickwick chintzes figured in linen-drapers' windows ... Boz cabs might be seen rattling through the streets, and the portrait of the author ... in the omnibusses ... a new and decidedly original genius had sprung up ... [with] little, if anything, in common with the novelists and essayists of the last century.'[41] This absorption by the public is long past and largely forgotten. Yet, *Pickwick*'s own success at widely transforming people's everyday perception may in itself partly impede the recovery of its real contribution for later readers, such as ourselves, for whom the book may now almost invisibly assume a reality it helped to generate.

In weighing this historic significance, *Pickwick* clearly represents only a small moment in the larger sweep of passenger transport history that provoked Dickens into awe at 'the South Eastern [Railway] Company for realising the Arabian Nights in these prose days' as it 'flew' him to Paris,[42] and that, after all, has made it possible to fly literally through the air to meet a person on the other side of the globe, traveling there in a matter of hours at speeds ten times faster than Dickens's train, to say nothing of those few who skyrocket into space at speeds more than thirty times faster than the airplanes—reaching in eight and a half minutes an orbit of 17,500 miles per hour. Still, by yoking the novel to the passenger transport system in *Pickwick*, Dickens hailed for his readers the networked community's coming together in their time. For author and reader alike, *Pickwick* thus represented a beginning that, having been accomplished, needn't be repeated. It set up for what was to come: fourteen novels all fundamentally different from *Pickwick*, where, now entering into the purposes of his characters and teaching his readers to feel their manifold multiplotted interconnectedness, Dickens inducted them into a world transformed by a public transport network built up initially around stage coaches and continuing to evolve all around.

III. Serialization

Much of *Pickwick*'s remaining renown comes from its place in publishing history as the novel that popularized serialization in monthly parts as a Victorian mode of writing, publishing, and reading novels. But what, if anything, does a revolution in passenger transportation have to do with the serialization of novels?

In one sense, the answer might seem obvious. This novel's serialization fits neatly into my preceding explication of how *Pickwick* intersects with a public transport revolution that networked people's living bodies together, merging their movement in shared time and space. After all, the ongoing monthly production of *Pickwick* also manifestly bound together author and readers as contemporaries around the serial novel, especially once Dickens subjoined the story to the passing months. Its serialization tied the novel to the here-and-now of the readers' living bodies, albeit via the communication system, and it lent the fictional characters a similarly irreversible past and semi-open-ended future that specially set them in contemporaneous time as well.[43]

Readers can still catch some of this contemporizing force reverberating within the story. For instance, at the point in its serialization when both Dickens and his readers had just finished three monthly numbers and a bit on the Fleet prison episode, Sam declares as he speeds off by coach:

'I wish them horses had been three months and better in the Fleet, Sir.'

'Why, Sam?' inquired Mr. Pickwick.

'Vy, Sir,' exclaimed Mr. Weller, rubbing his hands, 'how they would go if they had been!' (17.507)

In this seemingly straightforward example of playfully linking the story time to its monthly serialization, Sam's affiliating declaration was actually smoothing over and retroactively tying everyone temporally back together. There had been a lapse in monthly publication in June 1837, due to Mary Hogarth's death. Or—for another, more cerebral example—think of Dickens's serial readers, who truly had to forgo together foreknowledge of a story not yet written, contemplating the following piece of cognitive realism: 'Sam looked after [his father], till he turned a corner of the road, and then set forward on his walk to London. He meditated at first on the probable consequences of his own advice. . . . He dismissed the subject from his mind, however, with the consolatory reflection that time alone would shew; and this is just the reflection we would impress upon the reader' (10.282).

What's interesting about such minor examples of the novel playing backward and forward on its serialization is not merely that they show it impinging on its contents. Rather, in conjuring us imaginatively back to those first serial readers, such interplays also indicate the plain but unintuitive fact that Dickens engaged with his novel's ongoing monthly publication as it was happening. This novel's contents also framed its serialization for its readers. Or, perhaps better said, *Pickwick's* popularization of serialization neither preceded nor followed, but emerged in tandem with its story. As Mark Turner has recently said in an exceptionally smart discussion of serialization, Dickens 'thought about his novels *through* the serial form'.[44] As I will show, such a recognition ultimately undoes the currently prevailing story of *Pickwick's* serialization. That story—as Turner neatly recaps but leaves unchallenged—holds that 'the overall shape of *Pickwick* was haphazard and unplanned, and its unparalleled reception by the reading public a wholly accidental phenomenon'.[45] Both these claims will need to be revised. Against the usual one-way concern with how the serial mode of publication shapes a novel's contents, I will argue, as I have already conscripted Sam to hint, that Dickens's decision to structure his first novel around a regional public transport system gave it a form that endowed a newly communal meaning upon the serialization of novels, which, in turn, helps explain its popularization of serialization. At stake was not merely a publication practice, but also a serial form of knowing by which the networked community *Pickwick* depicted could, and would, be understood.

The serial publication of literature was not at all new in 1836, and focusing on *Pickwick*'s serializing of a series of stage-coach journeys around Britain immediately brings into the spotlight one serial predecessor to *Pickwick* that has habitually been relegated to scholarly footnotes: William Combe's *The Schoolmaster's Tour*. Combe's hugely popular verse narrative first appeared monthly in the *Poetical Magazine* from 1810 to 1812. It was subsequently extended with the publication of a 'second tour', issued in eight monthly parts in 1820, and then again in a third and final tour in 1821, also issued in eight monthly parts, after which *The Three Tours of Dr. Syntax* appeared in a pocket edition in 1826—the whole tours much reprinted, pirated, dramatized, imitated, and so on. Notably, George Cruikshank, who would illustrate *Sketches by 'Boz'*, contributed to one 1821 spin-off, *The Tour of Doctor Prosody*.

Combe's story relates the episodic misadventures of the good-natured, somewhat self-important hero Doctor Syntax as he rides about and around Britain on his horse Grizzle. It is classically picaresque. Doctor Syntax gets swept into a military exercise, mistakes a gentleman's house for an inn, visits a boarding school for young ladies, and has dozens of other adventures. One would not want, however, to emphasize the similarities of this story's events to *Pickwick*'s, as has been attempted.[46] Nor could their narrative styles be more different. The striking thing is the mode of production conjoined to Syntax's story's journeying structure. Here is Combe describing how he produced his tale on the fly for his engraver Rowlandson's illustrations, sounding for all the world like a versifying Dickens who never took over his story's reins from his illustrator: 'An Etching or a Drawing was accordingly sent to me every month, and I composed a certain proportion of pages in verse...and in this manner...the Artist continued designing, and I continued writing, every month for two years, 'till a work, containing near ten thousand Lines was produced.'[47]

Moreover, if one continues to look into this source, one also discovers that the Doctor Syntax famed in the 1810s and 1820s might

not necessarily have merely been the stuff of childhood memories for
the 24-year-old Dickens in 1836. Just two years before Chapman and
Hall presented Dickens with the proposition originating *Pickwick* that
he write some text monthly to accompany engravings done by the
famous illustrator Robert Seymour, a political parody appeared of
Lord Brougham as an updated Doctor Syntax traveling again around
England. It was called *The Schoolmaster Abroad* (London: T. McLean,
1834), and its author-illustrator was none other than Robert Seymour.[48]
Seymour's opening image is worth a look. It forwards the equestrian
figure of Doctor Syntax into a modernized passenger transport system (see Fig. 7).

So much for Syntax's horse Grizzle. Where William Combe's tours of
Doctor Syntax remind us that the ongoing monthly publication of a
series of journeys around Britain did not represent for Dickens any sort
of new departure, Robert Seymour's update hints at what would be
new: reshaping those serialized tours around a rapidly modernizing
public transport system. Very modern: Seymour includes a passenger
steam coach (in the engraving's lower right corner), and, as I mentioned

Figure 7. Robert Seymour, from *The Schoolmaster Abroad* (London: T. McLean,
1834). Courtesy of The Huntington Library.

in the Introduction, in the 1830s such steam coaches really did briefly exist, vying with horse-drawn stage coaches on the roads before capitulating, like the steam boats on the canals, to the railways and to the opposition mounted to their introduction by the entrenched coaching system.[49] By contrast Seymour invents Syntax's motorized trike. That solo-vehicle enables him to keep his hero firmly within the picaresque.

There was as yet a genre-breaking insight—a consilience—to come together. With hindsight, one can see it incubating in Dickens's first book, the collected *Sketches by 'Boz'* (1836). *Sketches* was published right before *Pickwick*, and in it, as J. Hillis Miller has brilliantly shown, Dickens self-consciously works toward crafting his 'Fiction of Realism' (the title of Miller's essay).[50] Amid a plethora of depictions of mobility, three moments especially grappled with the formal literary possibilities presented by the revolution in public transport.

First, there was the preface. When Dickens faced the challenge of articulating the coherence of the collected sketches a week or so before Chapman and Hall's proposal, he compared his book's project to a ballooning expedition: 'In humble imitation of a prudent course, universally adopted by aeronauts, the Author of these volumes throws them up as his pilot balloon, trusting it may catch some favourable current, and devoutly and earnestly hoping it may *go off well*' (p. iii). Having launched this metaphoric pilot balloon in his opening sentence, Dickens subsequently awkwardly turns out to have also embarked in it. This, however, allows him to extend his metaphor to the professional ascent he jocularly and humbly—balloons crashed a lot—confesses he pursues. As Cruikshank later flourished in a frontispiece to *Sketches* (Fig. 8), this ballooning metaphor positions the author and illustrator above a landscape they traverse widely. It thus provides a lofty way of embracing the scattered contents and relocating their coherence to the authorial aeronauts' exalted panoramic viewpoint.

In offering the balloon as a cynosure for *Sketches'* mobile eyeing of a mobile English world, Dickens imagines his book's unification not merely in terms of an amusing pastime but, as unlikely as it may seem today, public transport. In the 1830s ballooning looked potentially like the future. Piloted balloon ascents had begun in the 1780s. The Channel had been crossed in 1785. By Dickens's time, not only were there regular recreational ascents at Vauxhall, but pilot Charles Green had flown more than 200 experimental journeys. As one 1834 *Times* head-

Figure 8. Frontispiece to *Sketches by 'Boz'*, 20 monthly parts (London: Chapman and Hall, 1837–9), part 20. Courtesy of the Department of Special Collections, Young Research Library, UCLA.

line rallied, at a time when coaches were familiarly called 'machines', balloons looked like potential futuristic 'Aerial Machines'.[51] (Calling them 'flying machines' would have been confusing because the participle had been widely applied to any expedited passenger service— even fly-boats on canals.) That same year (1834) a European Aeronautical Society over-optimistically planned to book regular passage between London and Paris, and they opened a 'dockyard' on York Road and an office in Soho, rightly called by transport historian L. T. C. Rolt the planet's first airline ticket agency.[52] Dickens's figurative balloon was even especially timely. Later in 1836, much public excitement sur-rounded Charles Green's flight from England to central Europe, in which he reached Germany in a dazzling 18 hours. Dickens's long-time employer, the *Morning Chronicle*, carrying some of Dickens's sec-ond series of sketches, published a long account of the trip by a passenger, Monck Mason (21 December 1836), later reprinted as part of a little book Dickens owned, *Aeronautica; or Sketches illustrative of the Theory and Practice of Aerostation* (1838).

If Dickens's balloon framed *Sketches* in a way suggestively prefigur-ing *Pickwick*, it also, however, limned just what *Sketches* lacked. After all, the balloon's panoramic viewpoint by definition encompasses a lot indiscriminately. By contrast, in the stage-coach system passengers were channeled through a standardized space designed to network the variety of their activity.

Another moment in *Sketches* captured that aspect and offered something like an inchoate memorandum for forming *Pickwick*. At the end of 'Hackney Coach Stands', Dickens speculates that 'The autobiography of a broken-down hackney-coach would surely be as amusing as the autobiography of a broken-down hacknied dramatist. How many stories might be related of the different people it had conveyed on matters of business or profit—pleasure or pain! And how many melancholy tales of the same people at different periods! The country-girl—the showy, overdressed woman—the drunken prostitute! The raw apprentice—the dissipated spendthrift—the thief!' (1.231). Embryonic preludes of *Pickwick* spring into view here. There is the coach punningly imagined as a means to allow the end-less reintroduction of 'hackneyed' Hogarthian stories. There is the comparison between a coach, i.e. 'Pickwick', and a broken-down dramatist, i.e. 'Jingle', in conjunction with Jingle's dashed narrative 'stenography', as Pickwick will call it (3.68). And, pre-logically enough, in the next paragraph Dickens elaborates upon the accelera-tion of the system, which will be so important to *Pickwick*, contrast-ing the fast new cabs with the old, slow hackney coaches, 'coming to a—stand', as Dickens puns in closing.

Here again, however, the example also tells for its deficiency. Cabs and hackney-coaches both served short-distance, city-based public transport for individualized movement. Not only did they intensify the city as the proximate space of community, but also they lacked the key ingredient of the public transport system in which the individual

enters the system and becomes part of it through merging their journey with others.

All the ingredients really do begin to come together in a sketch
titled 'Early Coaches', which I mentioned previously. This sketch opens
with a paragraph decrying the coaches of earlier history as literally
equivalent to torture. Then, shifting 'early' to mean the present-day
misery of a stage coach departing at the crack of dawn in winter, it
recounts in detail the tribulations of suddenly being called upon to
take an impromptu journey, booking one's place, and heading out of
London. Prequel-like, the sketch thus first opens up the history of
stage coaching and then takes us right to the brink of that fast-driving
stage-coach journeying upon which *Pickwick* will embark.

And I really do mean, takes 'us'. Throughout the sketch, in a literary
technique variously experimented with elsewhere in *Sketches*, Dickens's
narration runs together the observing narrator 'Boz' with the reader in
a generalized second-person 'you' syntax. This 'you' is similar in many
ways to a third-person 'one'. The 'you' acts in a present-tense that at
once refers both to Boz's uniquely specific past present-moment and
also to some anyone's experience anytime, inclusive of the reader's.
Thus: 'Who has not experienced the miseries inevitably consequent
upon a summons to undertake a hasty journey? You receive an intima-
tion . . . that it will be necessary for you to leave town without delay . . .'
(2.172). And, to quote unarbitrarily: 'You left strict orders, overnight,
to be called at half-past four, and you have done nothing all night but
doze for five minutes at a time, and start up from a terrific dream of a
large church-clock with the small hand running round, with astonishing rapidity, to every figure on the dial-plate' (2.175).

In this you-form, Boz sharply observes the tension between his own
sense of venturing forth on his individual journey and his—and his
journey's—thoroughgoing interchangeability in the system: 'you,
yourself, with a feeling of dignity which you cannot altogether conceal, sally forth to the booking-office to secure your place. Here a
painful consciousness of your own unimportance first rushes on your
mind—the people are as cool and collected as if nobody were going
out of town, or as if a journey of a hundred odd miles were a mere
nothing' (2.172–3). In this moment, the stage-coach office induces a
switch of perspective that reduplicates in its content the sketch's overarching generalized 'you' form: the encounter mirrors the mode in
which Boz performs both as his inimitable self, utterly attuned to the

Figure 9. Dickens booking a stage coach. Illustration for 'Early Coaches' in *Sketches by 'Boz'*, 20 monthly parts (London: Chapman and Hall, 1837–9), part 7 (May 1838). Courtesy of the Department of Special Collections, Young Research Library, UCLA.

specific events befalling him, and as a common anybody, to whom those same events might well occur. The sketch's literary form thus finds its match in stage coaching's formal relations where individuals take their unique places in a system in which they are also understood nonetheless to be interchangeable.[53] After the publication of *Pickwick*, Cruikshank pinpointed this booking-office scene in 'Early Coaches' for illustration (Fig. 9).

Here Cruikshank depicts a clerk booking Dickens—'Boz'—for his journey in a book in the booking office. In a well-worn trope, the imaginary protagonist Boz thus gets constituted as real in relation to a book—as his book's readers really are. In this instance, however, that redoubling—in which Cruikshank inscribes in a real book an image of Boz's name being inscribed by a clerk in a book—specifies Dickens booking into the public transport system. Cruikshank's picture thereby presents more than merely a historical image of a stage-coaching office circa 1836. It suggests one artist's canny and clear-sighted commentary on a mode of realism produced by another.[54] It retrospectively discerns the departure to come. It insinuates *Pickwick* was never the accidental, haphazard upshot of pressure to serialize a story.

Sketches was coincidentally published on Monday, 8 February 1836, the very day Chapman and Hall made their proposal to Dickens for a monthly serial, and what Dickens faced on that day was therefore a new challenge to himself as an author. The shift from the collected *Sketches by 'Boz', Illustrative of Every-day Life, and Every-day People* to Chapman and Hall's serialized project presented a challenge of spatial and temporal concatenation. How to link together to 'form one tolerably harmonious whole' (1837 Preface, 19–20, p. viii) in an ongoing monthly publication his success at producing episodic tales, discrete scenes, and stand-alone characters, those 'little pictures of life and manners' (p. v), as he called them?

Here is Dickens describing what happened on that Monday, 8 February 1836, when William Hall paid him a visit and suggested to him that he join them in a serial publishing venture: 'The idea propounded to me was that the monthly something should be a vehicle for certain plates to be executed by Mr. Seymour, and there was a notion... that a "NIMROD Club," the members of which were to go out shooting, fishing and so forth, and getting themselves into difficulties... would be the best means of introducing these. I objected, on consideration, that... I was no great sportsman, *except in regard of all kinds of locomotion*; that the idea was not novel, and had been already much used... that I should like to take my own way... My views being deferred to, I thought of Mr. Pickwick, and wrote the first number' (1847 Preface, my emphasis).[55] Whether Dickens actually replied to Chapman and Hall's proposal by immediately redirecting the sporting club toward 'all kinds of locomotion', or whether 'on consideration' marks some small lapse in time, Dickens explicitly recounts substituting his own specialized transport content for the sporting-club genre proposed for the serialized format. One can hear his brain wave reverberating even in his description that 'a monthly something should be a vehicle'. A self-declared expert in locomotion, also up-and-coming fiction writer, thus 'thought of [a hero named for a fast-driving stage-coach line] and wrote the first number'.

Embarking on *Pickwick* represented a spatial and temporal transition from *Sketches*. In *Sketches* Dickens presented scenes and brief tales pre-dominantly based in and around London (set off against some sketches of 'our parish'). A 'speculative pedestrian' generally reports on this home turf, deeply attuned to the way his environs are infilled and abuzz with circulatory mobility in all kinds of forms (omnibuses and cabs, steam boats on the Thames, streets teeming with foot-passengers). By contrast, in *Pickwick* Dickens broadened to a perspective encom-passing inland regional road transport. He thereby also shifted the principle unifying the diversity of contents that constitutes *Sketches* from the spatial proximity of a single metropolis to a stage-coaching network. This shift in space was coupled with one in time. The discrete sketches first appeared as newspaper squibs, and when collected their characteristic mode still powerfully voiced a present-day journalistic immediacy in which each effervescent slice remains disconnected temporally from another. By integrating *Pickwick*'s monthly publica-tion with a protagonist's series of stage-coach journeys, not only did Dickens temporally concatenate the disconnected sketch form, but he also, equally importantly, cast his story as historically about the heyday of stage coaching nine years earlier and thus shifted the temporality of story and serial reader alike onto the shoulders of a public transport system that was engaging in constructing a common shared contem-poraneous time. Together these shifts in spatiotemporal narration are the break that *Pickwick* represented from *Sketches*.

From Dickens's authorial perspective, designing the novel as struc-tured by the stage-coaching system defined it in such a way as to embrace the eclectic ingredients and episodic plots that appeared to him at the time, as he straightforwardly tells us, an inevitable aspect of the serial mode of publication he faced. 'The publication of the book in monthly numbers', he recalled after completing the novel, 'rendered it an object of paramount importance that . . . the general design should be so simple as to sustain no injury from this desultory form of publi-cation.' The monthly format seemed to him at the time one in which 'it is obvious that no artfully interwoven or ingeniously complicated plot can with reason be expected' (19–20, pp. vii, viii). Instead, 'the author's object in this work, was to place before the reader a constant succession of characters and incidents' (19–20, p. vii). Dickens was, as he and his readers later discovered, dead wrong that ongoing serial production foreordained miscellaneous contents, episodic plotting, a fragmentary form. But no matter: in *Pickwick* he correctly saw himself as finding a way to give coherent meaning to his novel's ongoing serialization in such terms.

The long-standing and ubiquitous notion is thus that, as Mark Turner summarized, 'the overall shape of *Pickwick* was haphazard and unplanned' because that is how Dickens initially engaged with serialization. The problem is that this view misses a cardinal aspect of Dickens's achievement. What's striking is, rather, how specifically and sharply Dickens immediately carves his novel's overall generic shape. Besides in the title itself, perhaps nowhere is this clearer than in the advertisement that Dickens produced for *Pickwick* before its serial publication began.

In this advertisement, which was republished as a preface to the first American edition and is essentially a (periodicalized) preface to the novel, Dickens spells out his plan for the forthcoming serialized novel. In the ad's first two paragraphs Dickens sets out a spoof of 'Samuel Pickwick—the great traveller'.[56] Having announced that the Pickwick Club was established in 1822—lest his readers mistakenly think they really are returning to the days before the passenger transport revolution—the narrator proceeds to boast of Pickwick's 'celebrated journey to Birmingham', his 'penetrat[ion] to the very borders of Wales', and how the Pickwick Club's 'insatiable thirst for Travel' has led its members to explore 'the whole surface of Middlesex, a part of Surrey, a portion of Essex, and several square miles of Kent' (which is to say, London and its suburbs north-west and south-west, north-east and south-east). These heroes have even, the narrator facetiously informs us, 'navigated' the Thames in a steamer and 'fearlessly crossed the turbid Medway' (Rochester's river), two waterways packed with daily commuters.

Pickwick thus loudly departs from, and satirically collapses, a travel-literature genre in which strenuous voyaging carries protagonists far, spatially and temporally. In one unauthorized sequel to *Pickwick*, George Reynolds would send, as his title declares, *Pickwick Abroad* (1839), recounting, as its subtitle advertises, *The Tour in France*. But not only is that title unintentionally oxymoronic (a Bath stage-coach line for travel abroad?), but also lost, before one even begins to suffer through Reynolds's lame rip-off, is half of *Pickwick's* structuring jest. Where travel literature relied on the adventurous difference that moving individually across arduous distances could produce (such as one finds on the cover of another misconceived spin-off, *Pickwick in America*, where a Native American canoes Pickwick past a crocodile), the comparatively smooth and swift functioning of the modernized public transport system made a joke of travel as 'travail' and of the form of travel literature through which these *Pickwick* sequels were trying to forward Pickwick's journeying. The whole point was that the very transport that individuals had used to traverse from one place to another across time and space was inverted in *Pickwick* through a network creating a shared here-and-now of individuals in constant circulation.

After this genre-bending travesty of Pickwick as 'the great traveller', Dickens then veers into announcing the fiction of Boz and Seymour editing and illustrating a 'carefully preserved' Pickwick Club's archives that they will 'hand down to posterity'. This is the satiric subject of the advertisement's third and fourth paragraphs, balanced against the first two especially neatly in versions of the original ad that set them side-by-side in two columns. Now Dickens comically anoints his forthcoming work as the descendant of a noble historiographical lineage—the follow-on, he proclaims, to Edward Gibbon's *Decline and Fall of the Roman Empire*, David Hume's *A General History of Scotland*, and William Napier's *History of the War in the Peninsula*.

This facetious historiographic turn to the archive of *The Posthumous Papers of the Pickwick Club* does not, as it might seem, open up some second, unrelated avenue of parody for the novel: here is a travel story (ha ha!) that is also a trove of historical documents (ho ho!). As we saw previously, the *history* of the accelerating of passenger transport buttresses the spoof of travel literature and vice versa. Passenger transport's fantastically rapid transformations—first coaches, then trains—seriously make a joke of using it to look back to see people as historical contemporaries, while its swift and systematic interconnecting of people across distances was helping to create the very conditions of their communal temporal simultaneity.

So: 'Haphazard and unplanned'? Through this prefatorial advertisement's presentation of two linked genre counterstatements—reconfiguring travelogues and historical narratives—Dickens sharply stakes out at its outset the scope and shape of the stage-coaching novel his readers are going to read. He thus provides the signifying framework for understanding 'the series [that] will be completed in about twenty numbers', as the ad's fifth, and final, one-sentence paragraph announces.

The critical misunderstanding of *Pickwick* stems from the fact that the genre Dickens proposes calls for all kinds of subsidiary eclectic generic contents. As I explained previously, a crucial aspect that Dickens is going to show about the public transport system is that it carries and joins people together spatially and temporally whatever their separate purposes, whatever kinds of adventures they happen to be having in its terrain. This is why the ad also promises sweepingly to show 'public conveyances and their passengers, first-rate inns and road-side public houses, races, fairs, regattas, elections, meetings, market days—all the scenes that can possibly occur...' in its travel half, and, in its historical half, it bills the 'voluminous Transactions' as more heterogeneous work from the 'author of "Sketches Illustrative of Every Day Life, and Every Day People"'. Reading such promises, an editor like James Kinsley is not wrong to call the advertisement 'an open programme for a serial miscellany' (p. xxi). But he is mistaken in only doing so.

This novel does not have some secret happy unifying principle that rescues it from its eclecticism. This novel has a unifying principle that requires its eclecticism. The whole tired notion that *Pickwick* forces an evaluative decision between its formless heterogeneity and artistic unity is a false one, resulting in much unfortunate confusion. *Pickwick* makes nonsense of the idea of rolling together, as if they were one, its 'lack of cohesion and a clearly sustained single, structured vision' (to quote Mark Turner again).[57]

Instead, by crafting *Pickwick*'s genre around a public transport system that endowed a meaning upon the novel's eclectic and episodic contents, such that its presentation of a 'constant succession of characters and incidents' showcased how the public transport system unified people as a collective in shared space and time whatever the various activities happening in its reach, Dickens made his novel's serial publication format meaningful in forming a contemporaneous collective of its readers.

This is not to imply that fiction's serialization before *Pickwick* had no meaning. On the contrary, as Thomas Keymer has indisputably shown, authors in the eighteenth century, especially Sterne in *Tristram Shandy*, but also Richardson and others, tied the ongoing publication of their

novels in serial parts to both their authorial and their fictional protagonists' individual bodies. 'Serializing a Self', 'Dying by Numbers'—these
are Keymer's headings.[58] They do not, however, fit *Pickwick*. Rather the
picture that Keymer paints of eighteenth-century serialization—as oriented to the diary and memoir and yoking together through their serial
format the lone individual author's or protagonist's unfolding diurnal
life with the reader's—makes it all the clearer that Dickens's formal
structural use of the public transport system to endow meaning on his
story's eclectic contents made serialization newly matter because it
made, as his novel made, the *collective* of individuals meaningfully into
contemporaries in a shared network.

No wonder *Pickwick* miraculously drew together its audience from
across readers (and listeners) of a print market otherwise generally
highly segregated by class, politics, gender, and so on. No wonder it
was all the rage, as Mary Russell Mitford observed, with everyone
from 'the boys in the streets' to proper ladies and doctors, judges, and
others 'who are of the highest taste'.[59] In *Pickwick*, Dickens provided
readers with an armature for understanding serialization in terms of
the realist novel's project of imagining society as the networking
together of the open-ended, collective unfolding of people's individual
fates.

And isn't this in the sweep of the history of the novel one of the
chief reasons to care about fiction's serialization in the first place?
What matters about the serialization of the nineteenth-century novel
is that it materializes through its story's serial delivery the novel's formal capacity to express individual fates collectively networked as they
proceed from a shaping past toward an unwritten future. That meaning
of serialization had to be created historically, and that is what *Pickwick*
accomplished. And in accomplishing this, far from merely leading to a
slew of imitative serialized coaching adventures, *Pickwick* thus ushered
in serialization—whether in monthly parts or magazines, fascicles or
feuilletons—as the format for a whole variety of realist novels.

As publishing historians have patiently been explaining for years,
Pickwick did not represent some kind of a technological breakthrough
in the serial mode of production. Though there has been a persistent
myth that this novel introduced monthly part serialization, partly
re-enforced by somewhat misleading claims Dickens makes in his 1847
preface, it is equally well known that Dickens did not on that February
day in 1836 confront a newly invented, or even really a highly unusual

publishing format. Every serious critical discussion qualifies the novel's originality in this regard, such that countering its origin as an originary moment has become a fully predictable twist in telling the story of the novel's serialization.[60] But it is not quite therefore that 'Dickens and his publishers discovered the potential of serial publication virtually by accident', as Robert Patten long ago concluded.[61] That is only how it appears when viewed solely as a publishing phenomenon.

Dickens's story made new sense of serialization for its readers. And opening up the story's internal clockwork at a key moment helps show how. Apprised by his lawyer that he will be taken to prison if he does not pay Dodson and Fogg, Pickwick asks two questions. The first is about timing: 'When can they do this?' The lawyer's answer put readers inside a legal chronology, erecting, in plot terms, a barrier time: 'just two months hence, my dear Sir'. In reply, Dickens then has Pickwick explicitly throw up against this legal alarm clock set to go off in two monthly numbers his own use of time: ' "Very good," said Mr. Pickwick. "Until that time, my dear fellow, let me hear no more of the matter. And now," continued Mr. Pickwick, looking round on his friends with a good-humoured smile . . . "the only question is, Where shall we go to next?" ' (13.371).

Characters, author, serial readers: all are imagined inside the time of a public transport system. By contrast with that oppressive legal calendar, repeatedly appointing the dates and places at which justice will not be served up, time becomes the free-wheeling accretion of ongoing activity happening synchronously within easy stage-coach reach. The bewilderment that this novel so often generates among readers as it adds and subtracts this or that secondary character as we fly along its road-tripping—now Snodgrass, Tupman, and Winkle, now Old Mr Wardle chasing after Jingle and Rachel, now Bob Sawyer and Ben Allen, a swirling 'series of adventures, in which the scenes are ever-changing, and the characters come and go like the men and women we encounter in the real world' (as Dickens described it in a preface, 19–20, p. viii)—is fully part of its inclusive network message. The question 'Where shall we go next?' poses no decision of plot. Its meaning lies in its indecisiveness, in its arbitrary selection of one possible stop in a regional transport system filled with individuals' ongoing activity. And, as it happens, this time the journey is to Bath. 'Sam was at once dispatched to the White Horse Cellar, to take five places by the half-past seven o'clock coach, next morning' (13.371); the crew are in Bath by precisely seven at night; they have met on the ride one of the main protagonists of their next adventure, Mr Dowler ('We are going to be fellow travellers...', 13.373); and, unforgettably, Sam has been outraged by the fact that 'The names is not only down on the vay-bill...but they've painted vun on 'em up, on the door o' the coach': 'the magic name of PICKWICK!' (13.374).

Whatever subsequently occurs in those four chapters in Bath, the disorderly openness of the adventures and the diversity of the new characters they meet there contribute to a construction of time as the extemporaneous, non-teleological concatenation of ongoing diverse activities. It may thus appear—it has been made to appear—as if somehow 'ordering in *Pickwick* is more or less synonymous with chronology'. Or, that Pickwick's question 'Where shall we go next?' is nothing but chronology, and the fragmented novel organized around an 'and then...and then...' syntax that mirrors the novel's serialization—'and then another number'.[62] But, as we have seen, that is not quite the whole truth. Whatever happens next there in Bath, Dickens makes those four chapters of adventures the chronicles of a public transport revolution whose schedules indicate its nearly continuous availability (against plotted barrier times) and whose scope and reach define a space of circulation ideally antithetical to blockages. Thus here 'we' go for 'two months' (never mind for now the difference between a ride and a read) such that *Pickwick* realizes for its serialization the meaningfulness of its forming a contemporary collective of its readers.

Fast forward past Winkle's sedan-chair debacle in Bath's Crescent Row and his subsequent stumbling across Bob Sawyer and Ben Allen while asking for some directions in Bristol. As two months and two monthly numbers lapse, the ticking of narrative time loudens as the legal clock prepares to disrupt, as promised, the ongoing serial stagecoaching story: 'Trinity Term commenced. On the expiration of its first week, Mr. Pickwick and his friends returned to London.... On the third morning after their arrival...a queer sort of fresh painted vehicle drove up.... The vehicle was not exactly a gig, neither was it a stanhope. It was not what is currently denominated a dog-cart, neither was it a taxed cart, nor a chaise-cart, nor a guillotined cabriolet; and yet it had something of the character of each and every of these machines. It was painted bright yellow, with the shafts and wheels picked out in black....' (14.425). What is it? It is a nineteenth-century police car, and it is pulling up in descriptive slow motion to pull over Pickwick's stage-coaching adventures. The arrest that follows will bring to a halt the continuous sequential compilation of eclectic happenings around the bustling intercity circulation of the transport network. Pickwick will seem to come to a stop for 'three months and better in the Fleet'. The serial *Pickwick* will continue publication, but (as was previously explained) the story's tarrying with an intensified, inverted formulation of community in the Fleet, where individuals are bounded by spatial proximity over time, makes Pickwick's release the serial's as well.

' "I wish them horses had been three months and better in the Fleet, Sir." "Why, Sam?" inquired Mr. Pickwick. "Vy, Sir," exclaimed Mr. Weller, rubbing his hands, "how they would go if they had been!" ' Here, as in the novel as a whole, Dickens reconstitutes his novel's serialization in terms of its stage-coaching structure. The effect is to read of 'a swarming plurality of isolated centers of vitality', a 'simultaneous plurality', as J. Hillis Miller nicely articulated it.[63] Dickens had begun to think the networked community through his novel's serialization.

IV. Systems

A public transport system interconnects and expands the here-and-now experienced by people's living bodies into its circulating network. As people move in that network, scheduling their swift intersecting 'nows' on spatial trajectories gives rise to, and cause for, a common convention synchronizing their time and relates the interconnected places to each other in simultaneous time, while their individual 'heres' outspread into and along the network collectively create a public space in which ongoing social activity, or inactivity, is seen to take place. I now want to turn to the significance of this aspect of the passenger transport system specifically as a system: that the here-and-now of mobile individuals is its contents, that it arrays living bodies into a network. Just mentioning this fact immediately raises, I think, the passenger transport system's historical and overlapping relationship with the postal system, and most of this concluding discussion of *Pickwick* is dedicated to considering what *Pickwick* says—it has a lot to say—about letters, the postal system, and epistolary form. My argument is essentially that this novel, as a novel structured as a series of rides around the regional public transport system, hammers home the historical reconfiguration of relations that the passenger transport revolution meant for the post and for the novel's epistolary tradition. Beyond this, by following out the implications of one letter in particular (the story's last), I will also suggest that one may distinguish the role that the public transport system plays as a system among systems generally, in *Pickwick*'s rosy view of it. For Dickens, opposing the passenger transport system as a system to other systems helped him to inaugurate his famous comic style in which he satirizes his contemporary middle-class administrative society as regulated by inhumane bureaucratic systems.[64]

First, though, the postal system and the public transport system: briefly, some comparative history. Those two systems—for letters and for passengers—both offering in their different ways to bridge people's physical separation, were newly brought together and revolutionized by John Palmer's introduction of Royal Mail coaches in 1784, (turn-pike) gates thrown open and (mail guards') bugles trumpeting their every arrival. Before Palmer's reform of the postal system in 1784, *to post* was, as the verb's two meanings indicate, both the fastest method of travel for a rider using stages to change his or her horse and also the mail courier service for which that staging system was originally created to speed the King's messages along with some privileged others' letters. While as part of the complexity of things coaches might sometimes be used to deliver mail, until Palmer's reform the letter post was based around mounted post boys with their letter bags, who were notoriously slow about their business. The same method—posting—was thus used both for letters and for people, but the two did not ordinarily move together. Only with the Royal Mail coaching system introduced by Palmer did the mail become the 'conveyance that outran every other', and then the few seats it offered to passengers drew humans into a system whose unprecedented speed also made it the cutting edge of a passenger transport system, despite its often nocturnal schedule.

During Dickens's life up to and through the writing of *Pickwick*, besides carrying letters the Royal Mail coaches thus also formed part of the passenger transport system, naturalizing, for instance, the otherwise bizarre phrase 'I am coming down by the next mail'. As Dickens was writing *Pickwick*, the railways were absorbing this double function. Steam trains had begun carrying mail in the 1830s, and, for example, in the summer of 1837, after the completion of the Grand Junction Railway, a rail carriage replaced the mail coach between

London and Birmingham. Only in the decades after the publication of *Pickwick*, with the advent of the electric telegraph, would the public communication system and the public transport system begin to diverge radically and visibly, taking separate routes at vastly unequal transmission speeds.

Looking back to this brief period when the two systems noticeably ran together, one can initially perceive that they both underwent an essentially similar radical transformation as the stage coaches speeded up. The whole point of Rowland Hill's breakthrough pamphlet *Post Office Reform*, published early in 1837, when Dickens was roughly halfway through *Pickwick*, was that, within the swift reach of the transport network, temporal and spatial distance clearly no longer counted as it had before and Britain should be treated as a single networked space of circulation with 'Uniform Penny Postage' (the title Hill first proposed for his pamphlet). At the time Hill announced his proposal for a national penny post virtually every town in the nation already had a regional penny post, and it is not hard to see, given that those cities were now closely and regularly connected by a fast long-distance stage-coach network, being replaced by an even faster railway network, that it made perfect sense for all those local penny posts to merge into one. In this regard, the mail and the passenger transport system look something like twin fronts in the networking of Britain. A swift postal system was, like the passenger transport system, annihilating separating spatial and temporal distances.[65]

The key point, however, which I want to suggest Dickens grasped in *Pickwick*, is that this mutual relation of the two systems, in which both were similarly transformed by the stage coaches' and then the railways' overcoming of separating distances, actually represented a fundamental transmuting of the historic function of the postal system: letters written in the context of the new public transport system now circulated around a network of people already understood to be networked in person. It was not only that letters came now to look something more like the messages sent between people moving around the same city—i.e. around a circulating 'here' defined by a physically proximate social space. Now it was not their proximate physical distance but another network—the passenger transport network—that connected these people in person. The mail, which by definition provides a means to communicate without the necessity of being bodily present, began also to operate in tandem with people's actual ordinary and continual bodily circulation around a passenger transport network, growing together with that circulation. 'Dear Tom,' Dickens writes in a typical letter, 'I arrived here (57 miles from Exeter) at 8 yesterday Evening....'

For the novel as a genre, from within, that is, the perspective of *Pickwick*'s stage coaching around the passenger transport network, letters thus lost a magical power they had once had—that an entire important epistolary tradition of the novel had once made theirs—to link people together in writing as if they were present to each other. Why would anyone now try to sustain such a discursive detour? That would look something like a ridiculous failure to recognize that letters do not actually transport living bodies. It would look, that is, like a sequence of farcical scenes in *Pickwick*, all of whose punch lines hinge upon how the postal system, unlike the passenger transport system, does not carry living bodies, which simple fact has now come to mean that a whole declining romantic eighteenth-century epistolary subgenre of the novel that once usefully created a literary convention conveying its readers into the contemporaneous here-and-now of a sub-network of characters—'writing to the moment', as Samuel Richardson famously described it—must really and truly give up the ghost.[66]

Alas, one might sigh, for love letters. In *Pickwick*, the reader only gets to fall for them as pieces of mail. As an opening move in his romance, Sam Weller will laboriously craft a valentine. Instead of involving the affairs of his hopeful heart, the whole of Sam's postal confection, prompted by an advertisement in a shop window and then subjected to 'Mr. Weller the elder deliver[ing] some Critical Sentiments respecting [its] Literary composition', presents a protracted examination of the letter's formal properties (e.g. 'the great art o' letter writin'' requiring leaving off so 'she'll vish there was more') and its material creation as a physical gift ('To ladies and gentlemen who are not in the habit of devoting themselves to the science of penmanship, writing a letter is no very easy task...') (12.339, 344, 341). And even with the reader thus transfixed by the ink-blotted and rhetorical construct of Sam's love letter, Dickens still manages to disrupt any reigning postal sense of order by having Sam nonchalantly sign his valentine, merely for poetic effect (in the only part actually typeset as a letter):

'Your love-sick
Pickwick.' (12.345)

There's a new way, open to us all, of fulfilling the epistolary convention proscribing 'sign[ing] a walentine with your own name' (12.344). Whatever other freedoms sending a valentine licenses, one would ordinarily not be likely to desire transferring its origin in one's living flirting body to another's, but Sam deliciously trifles with the post's need to stabilize identities always by definition asunder because his is the comic perspective of this coaching novel where letters create—are defined by—their spatial and temporal bodily disconnect.

In a previous century that disconnection generally seriously made, as it farcically makes here, for the sharpest effects of literary composition in a romantic novelistic epistolary tradition. That tradition not only seduced its readers through its letters' power to produce a sense of immediacy, making present both characters and story, but also delighted in pulling its readers up short with a consequential jolt, self-awarely reminding its readers that they were reading letters, which, for instance, might cross in transmission or end up in the wrong hands. In *Pickwick*, where the post undoubtedly connects people through their communications as sweetly sometimes as Sam's valentine, that same disconnection upon which the epistolary tradition so consciously founded itself reappears as ridiculous. In the context of a passenger transport revolution the post's expanding power to allow separated individuals to communicate with each other across time and space without bodily having to traverse that time and space has become, as never before, also its limitation to its not being coterminous with their living bodies. And the reason this matters, the reason Dickens will parodically slough off an eighteenth-century novelistic epistolary convention, is not that the post stands in some kind of secondary relation to the public transport system as a means of networking people together, or even that one or the other system is necessarily wholesale better or worse at constructing community or connection, but simply that *Pickwick* articulates *the passenger network* as a network linking together people in time and space.

And so as Sam carefully folds and labels his valentine with Mary's address, provided here in full, to make it 'ready for the General Post' (12.345), and then as he and his father casually begin talking about Pickwick's upcoming trial, specifically about how it requires an alibi defense, that is, appositely about the defense of being bodily elsewhere, one can only laugh to realize that Dickens is purposely amplifying exponentially the small worry about whether Mary will think Sam's valentine is from Pickwick by expounding a scene that underscores that Sam has potentially dangerously sent off this misattributed love letter just before Pickwick's trial for breach of promise. But Dickens is baiting his readers. He is schooling readers in a new realism of letter-writing, and so—it is part of Dickens's point—no repercussions will materialize from Sam's valentine. It floats away into the postal ether, never to be mentioned again.

Instead in the next chapter, materializing out of nowhere, come Pickwick's love letters. Or rather, at his trial, the lawyer Buzfuz introduces into evidence two hasty notes Pickwick has written to his housekeeper, which the barrister outrageously interprets as love letters: 'Dear Mrs. B.—Chops and Tomata sauce. Yours, Pickwick.' and 'Dear Mrs. B., I shall not be at home till to-morrow. Slow coach. Dont trouble yourself about the warming-pan' (12.359). To eat, to return home, and sleep in a warm bed: Pickwick's two undated messages are limited almost exclusively to relaying his body's trajectory in time and space, and their meaningfulness is essentially that they are virtually meaningless disconnected from his situated movement once past. Except, that is, in that maw of insidious competitive self-interested reinterpretation, the legal system: there the prosecuting barrister deciphers Pickwick's messages as if they were the secret correspondence of a kind of latter-day Lovelace, whose Clarissa were finally getting the breach-of-promise trial she always deserved, one that—sweet vengeance—will even legally imprison her seducer for a debt of 150 pounds, as she was. Farewell, Richardson. And not just *Clarissa*. Dickens surely counts here on Pickwick's supposed seduction of his landlady-housekeeper being read against that other housekeeper, Richardson's Pamela, and her landlord, 'Mr. B.'—'Dear *Mrs* B. . . .'

Shamela and *Joseph Andrews* are nothing to this. Those parodies of Richardson's epistolary romances by Fielding had no replacement to offer. Dickens has a passenger transport revolution behind him. Hence in his address to the jury the prosecuting barrister—Dickens mercilessly sets him up—trips over precisely that which ironizes his analysis of Pickwick's letters: 'And what does this allusion to the slow coach mean?' the prosecutor demands, 'For aught I know, it may be a reference to Pickwick himself, who has most unquestionably been a criminally slow coach during the whole of this transaction, but whose speed will now be very unexpectedly accelerated, and whose wheels, gentlemen, as he will find to his cost, will very soon be greased by you!' (12.359). Pickwick, a slow coach? Like Pickwick, whose silenced perspective frames the reader's reception of this trial, readers are meant to writhe in indignation. Slow coach?—just this one smirking line of the barrister's stentorian pseudo-analysis is twice as long as both of Pickwick's letters combined. Hearing this insult—hurled here in print for the first time according to the *OED*—some one of Dickens's London readers might have wished to hoot back to the lawyer's lame play on 'Pickwick' as a (new, fast-driving, intercity) coaching line that his own client, 'Bardell', was the name of a (slow) London hackney-coach service in the 1820s.[67]

Readers are inside an extended comedy routine, and the joke splitting in all directions, including recursively onto the novel itself, is that postal letters once composed and mailed enter into a temporal and spatial dimension all of their own, and one separate from that of living bodies. The fact that the fictional characters also lack living bodies and all that entails, particularly that they do not actually possess a here-and-now but that the realist novel produces in written language the reader's sense that they do by deploying language referencing a situational context as if a body were moving through it, is a discussion for another day (about deixis in the novel). The goal here is simply to avoid the tempting deconstructive word slide that the critic John Bowen makes when he argues that because characters may ride the mails or travel post, 'the characters . . . are . . . posting themselves'.[68]

Neither is it hard to disambiguate the two systems. Perhaps nowhere is the crucial incongruity between the transport and postal systems clearer than when Dickens has Pickwick for the first time undertake to deliver a letter (from Winkle asking his father's belated blessing on his marriage to a very un-Harlowlike Arabella). This trip to Birmingham, Pickwick's last journey (which I discussed previously), introduces the novel's final interpolated tale. This is the remarkable 'Story of the Bagman's Uncle', which recounts a ride in the ghost of an old Royal Mail coach by Jack Martin, pointedly described as 'the only living person who had ever been taken as a passenger on one of these excursions' (17.530). In this eighteenth-century dream tale of a romance kindling inside a Royal Mail coach, which yoked the historical contemporaneity of its characters to the speeding of the coach carrying them, Dickens also laughingly pries apart the Royal Mail coach's services as a passenger transport system carrying live bodies from its postal purposes. The fairytale's mock-moral, also punch line, is that living people cannot dream themselves back into the letter bags of those 'ghosts of mail-coaches'. As the unepistolary, alphabet-encoded novel winks twice, even if they could, all that they would find in those spectral mail pouches would be: 'The dead letters of course' (17.530).[69]

After this cautionary ghost tale—with its message reminiscent of the grim vignette of the man struck down by apoplexy who 'fell with his head in his own letter-box, and there he lay for eighteen months' because 'Every body thought he'd gone out of town' (8.212)—what else should one expect but that delivering the letter to Winkle's father in Birmingham will be a complete waste of time for the 'living' characters. The commingling of Winkle's written correspondence with Pickwick's personal delivery backfires staggeringly. Interference, disruption, boredom, tension: two incompatible systems made to collide. Upon their arrival at the senior Mr Winkle's, Bob Sawyer fidgets till he is caught making clowning faces and is promptly squelched into 'humility and confusion'. Pickwick nervously waits and waits and 'eye[s Mr Winkle] intently as he turn[s] from the bottom line of the first page to the top line of the second, and from the bottom of the second to the top of the third, and from the bottom of the third to the top of the fourth' (18.540). It is as if the fictional characters' living were temporarily suspended until the completion of that silent disembodied connection being procured by 'four closely written sides of extra superfine wire-wove penitence' (18.540), while Dickens's readers, reading of the reading of this letter, far from being aligned with the epistle's lector, find themselves positioned along with those traveling visitors looking from the outside at the lines of writing and sheets of paper, superfine and wire-woven.

'Just when Mr. Pickwick expected some great outbreak of feeling, [Mr Winkle] dipped a pen in the inkstand', asks his son's address, endorses it on the letter, and subsequently has 'nothing to say' (18.541). Locked out of that other systems' circuits, the would-be envoys turned couriers are, in their words, 'floored' (18.542). Only the return coach trip, though made in the most dismal rain and mud, revives them: 'there was something in the very motion and the sense of being up and doing' (18.543). 'Up and doing'? 'The very motion'? In a moment, I will have something to say about the way in which the coaching system's uptake and organization of our characters' living bodily movement across time and space into a network so easily collapses, as it does here, into simply being alive and moving. For now, though, I will keep tracking the letters; their implosive epistolary anticlimax nears.

Naturally Winkle's letter to his father about his elopement will not call forth a letter in response. Why should it? In the context of the passenger transport revolution depicted in *Pickwick*, that would be a ridiculous affectation, something savoring quaintly of the last century. *Pickwick* races past the likes of *Humphry Clinker* as if those tours around Britain told in letters were an old post-coach broken down by the wayside. *Pickwick* is an anti-epistolary novel, the story of a corresponding club that never did. A few days later Winkle's father will make the easy trip from Birmingham to London to talk with the couple himself. 'The passage is made in a very short time, at a very cheap rate, and without the least difficulty', Dickens would write in 1836 to his new wife Catherine's grandparents, inviting them down to London from Edinburgh, more than twice as far away as Birmingham.[70] Winkle's letter to his father about his new bride is like all the letters in *Pickwick*, even those introduced into the text in full, such as the letter from Dodson and Fogg declaring the commencement of their legal action that specifically requests a response by return post, or the one Sam receives in Bath from the insufferably snobby butler John Smauker inviting him to a 'swarry', or the one in which romantic Tracy Tupman declares that, crushed by the loss of Rachel Wardle, he has gone away never to be seen again. All elicit no further correspondence, just immediate visits.

In this meaningfully lopped epistolary mode, where Pickwick is always making round-trips and letters aren't, by the close of the Birmingham journey, which is Pickwick's last trip, Dickens has virtually methodically given a complete sequence beginning with the composition of a love-letter for the mail (Sam's valentine), followed up by extended episodes concerning such letters as they float free of bodies (the lawyer on Pickwick's letters) and are transmitted (Jack Martin's ride on the Royal Mail coach) and then delivered (Winkle's father). Composition, transmission, delivery... Sam finds upon returning from Pickwick's very last journey that 'There's a letter been waiting here for you four days; you hadn't been gone away half an hour when it came; and more than that, it's got, immediate, on the outside' (18.554).

It should come as no surprise that this final scene in which Sam receives and reads a letter from his father, an eye-opening act silently shared with us, makes good structural sense as a final pronounced act of reception rounding out Dickens's step-by-step dismantling of epistolary form. But more than this, Mr Weller's letter to Sam also closes

the novel's meaningful travesty of epistolarity with a startling enfolding
into its contents of the salient difference in form between the postal
system and the public transport system—that the one carries written
communications, the other living bodies. Its subject is the dying of
Mrs Weller. Not only is how her death gets figured remarkable but also
its figuration contributes to the way Mr Weller's letter constitutes itself
as an event—brace yourself for it—of letter writing and reading:

> '*Markis Gran*
> '*By dorken*
> '*Wens*^{dy.}

'My dear Sammle,
'I am wery sorry to have the plessure of bein a Bear of ill news your Mother
in law cort cold consekens of imprudently settin too long on the damp grass
in the rain . . . the doctor says that if she'd svallo'd varm brandy and vater arter-
vards instead of afore she mightn't have been no vus her veels wos immedetly
greased and everythink done to set her a goin as could be inwented your
farther had hopes as she vould have vorked around as usual but just as she wos
a turnen the corner my boy she took the wrong road and vent down hill vith
a welocity you never see and notvithstandin that the drag was put on drectly
by the medikel man it wornt of no use at all for she paid the last pike at twenty
minutes afore six o'clock yesterday evenin havin done the journy wery much
under the reglar time vich praps was partly owen to her haven taken in wery
little luggage by the vay your father says that if you vill come and see me
Sammy he vill take it as a wery great favor for I am wery lonely Samivel N. B
he *vill* have it spelt that vay, vich I say ant right and as there is sich a many
things to settle he is sure your guvner [Pickwick] wont object of course he vill
not Sammy for I knows him better so he sends his dooty in vhich I join and
am Samivel infernally yours
> 'TONY VELLER.' (18.554–5)

Now there is some writing to the moment. Dickens brilliantly fails here
to perform the temporal special effects and embedded contextual refer-
ences ordinarily required for epistolary exchanges to present themselves
transparently as connecting communications. Spelling debates? Punctu-
ation? Pronouns? 'Wot a incomprehensible letter,' Sam declares. 'Who's
to know wot it means vith all this he-ing and I-ing!' Sam can only begin
to decode its message by grasping that in the streaming moments of its
composition, Mr Weller 'got somebody to write it for him' and
'then . . . comes a lookin' over him, and complicates the whole concern'
(18.555). In terms of the mini-farce upon epistolarity *Pickwick* is perpe-

trating, this letter presents something like the explanatory apogee of Dickens's reinterpretative comedy of letters as a form of distal communication—and a structural endpoint. Does Mr Weller's letter, thrusting us back to the scene of its bodily production, require Sam and the reader to envision Mr Weller's physically composing it live along with someone else? That will not be too hard: Dickens has offered up that scenario before. Only then Sam was writing a valentine and Mr Weller providing the literary advice, and it was the opening of a romantic letter-writing process, where it is here, with the death of Mrs Weller, the close.

Meanwhile, one can see Mr Weller's letter is not a dizzying jumble, but a meaningful one, even the *pièce de résistance* of the epistolary sequence that I have been tracing, because inseparable from the elucidatory unsorting of the pronouns and verb tenses, whose suppression would ordinarily be required to create a letter reader's illusion of being directly addressed by another person, comes an extended coaching metaphor that further countermines the whole epistolary form by tying coaching to the immediate presence of living—dying—bodies. Far from being a bolt of ill news that arrives as a surprise immediately after the last stage-coach journey ends, Weller's letter represents a meaningful conclusion to the last stage-coach journey, even to all the journeys, as these repeatedly reframe the postal network and the epistolary history of the novel from a standpoint established by a revolution in public transport. From that standpoint, the stage coach in Weller's letter provides much more than a convenient metaphor for a living and dying body. Life and death take on new meanings in the new historical context, and the new meaning specifically being conveyed here is, once again, that stage coaching's networking of living bodies makes newly significant the postal system's not doing so and comically collapses a literary epistolary tradition that self-consciously used this difference between written messages and living bodies to construct a fictional form in which letters provided the means to portray the networking of people across time and space.

Mr Weller's letter also offers, for the first time, a reverse-angle view of the stage-coaching story as it appears from the perspective of epistolarity. By having Mr Weller figure Mrs Weller's dying as a stage-coach ride, Dickens folds the formal relation of passenger stage coach to letter into a letter's content. To put it tendentiously, Weller's letter represents an author figuring stage coaching from within a letter. Instead of merely the stage-coaching novel framing epistolarity, epistolarity momentarily also enframes the stage-coach story framing it. Describing it this way makes it sound more supersubtle than it is, and what the switch—Weller's figuring of his wife's dying as a stage-coach ride in a letter—quite straightforwardly brings into view is that the stage coaching in *Pickwick* plays at the same game the epistolary novel did, self-awarely convenes around the same literary convention. The passenger transport network also works to create the illusion of the immediacy of its characters' presence and present. It too constructs the imaginary literary characters—lacking, as they are, real living bodies—as immediately present to each other and to the reader, such that their living presence, or dying, can collapse metaphorically not into their letter writing but into their stage-coaching movement.

Within Mr Weller's letter, the stage coach flattens ahistorically into life as endless motion. Applying historicizing pressure gets one nowhere. At best it reinforces the idea that Old Weller seems merely to be comically revivifying and updating death's own dead metaphor as reaching the end of one's road. His figuring of the body's dying as a stage-coach journey gone awry thus signals truths that would seem to have ever been true: that forkings in the road fix fates, that life speeds along in relation to a clock—Susan Weller's, the letter notes in a sad detail, crossing the ultimate public toll-gate at 'twenty minutes afore six'—and even that being thin 'as a rail' may laughably mean not having taken on any extra luggage. It does not really matter whether Dickens is counting here on his readers hearing echoes of familiar old truisms about 'Life's uncertain voyage' (Shakespeare) or whether he is fabricating a new cliché his readers retrospectively project and discover as eternally true (gliding, for instance, over the difference between a voyage and a regional intercity trip with turnpikes and so on). Either way, the historical force of the contemporary transport revolution sinks away as Susan Weller's dying and its figuration as stage-coaching movement collapses her life into its movement.

The reason this matters is that in Mr Weller's letter, partly because it is a letter, *Pickwick* confronts starkly a pervasive mystification that arises from the facility with which passenger transport gets merely equated with activity and life, 'the very motion'. Life thus appears eternally figural as a road, humans forever mobile in space and time 'all the voyage of their life' (Shakespeare again). The effect is to cloak the historical specificity of the rise of the public transport system in ubiquitous timeless trueness. In this view, it seems overreaching to historicize Pickwick's question 'Where shall we go to next?' or his 'going'. The stage coach vanishes into scenic period detail. It's only human creatures on their roads, as always. Understanding this mystification cuts, moreover, to the very existence of this study. What after all, one might fairly have demanded, had produced the circumstances in which aspects of the historical import of the public transport revolution, and Dickens's insights about it, had been overlooked? Isn't the argument partly that this passenger transport history is palpably present in Dickens? Previously it was only suggested that the undeniably momentous acceleration of indefatigable, iron-wrought, steam-driven locomotives had produced a mistakenly nostalgic view of stage coaching that misinterpreted it right into a pre-industrial agrarian grave. Now it may be added that there has also been a quick and easy metaphoric slip that has allowed the stage-coaching story, despite its being sharply pegged to a historical revolution in public transport specific to its time, to go unremarked because it also confusingly appears ahistorically as if it were merely vividly updating a truism of living bodies ever on the move. (Such a combination is not as contradictory as it might appear. It presents a quintessential feature of much ideological work: to naturalize and to enshrine as self-evidently true contingently historical social relations—in this case, those contrived by a public transport revolution.)

'N. B' (*sic*, lacking a period after the 'B'), note well: missing and mispunctuation, spelling debates, wayward grammar, all are not incidental to the send up of epistolarity but utterly relevant, and what they bespeak is that letters may not carry living bodies but that the writing of letters as a means to writing a novel has, unlike any coach, the tremendous benefit of operating in the same medium as the novel. In this case, for instance, the fact that letter writing stores writing in a written novel that readers, inside the story and out, are reading makes for the whole delightful inseparability of our reading from Sam's reading from Weller's composition, all the temporally separate moments coming simultaneously to eye. In the toils of an epistolary novel, the realist illusion that one is entering into a shared here-and-now with the story's diegetic characters arises partly from the act of participating in the same activity with them—reading writing—and thus not despite, but because, that activity denotes their spatial and temporal disconnection. The convention through which published, printed, typeset words, bound in books, get read as if they were handwritten missives further allows the reader to take in this indexicality of letters (pointing to reading and writing persons) in sustaining contradiction with the knowledge that a novelist has imaginatively produced and arranged it all. This alloy creates the experience of fictional realism, allowing the reader to clinch together two irreconcilable positions that, precisely in being held together, generate the form's meaningfulness.[71] Through, however, the pronounced illiteracy of this particular epistolary *trompe l'œil*, Dickens subverts from below—'infernally yours'!—that once-happy realist circle, constituted around the literacy of imaginary people writing letters and novel readers reading them.

But notice also that Mr Weller's letter does not merely forcibly evacuate epistolary form. It equally and blatantly supplants standards of literacy generally: 'N. B he *vill* have it spelt that vay, vich I say ant right.' (One special marker of this disruption—which continually erupts throughout the novel—is the irresolvable tension over whether the family name should be spelled with a 'W' or a 'V.') And just here Weller's letter can also offer some further insight about this novel's handling of language and systems more generally.

Consider again the letter's first few words: 'I am wery sorry to have the plessure of bein a Bear of ill news...'. In most any other early nineteenth-century novel such a line would signal the stock character of an illiterate buffoon. Especially in key near-predecessors to *Pickwick* like Robert Surtees's *Jorrocks* or Pierce Egan's *Life in London*, that opening line, along with the entirety of the letter that followed, would read as an author crafting illuminating malapropisms at his humble character's expense, however humbly beloved be the character. Dickens handles it differently. Whenever challenged, Mr Weller always proves fully aware of his ostensible semantic and grammatical mistakes, and readers readily perceive that this character holds to his insightful linguistic glitches as making his meanings. As he asserts, 'I know wot's o'clock, Sir. Wen I don't I'll ask you, Sir' (15.465); he 'didn't drive a coach for so many years, not to be ekal to his own langvidge'.[72] Hence, if anything, it can seem that Mr Weller, like Dickens, commands the standard spelling or the appropriate word or the proper grammar, and that both dispense with the drably correct in favor of dazzling phonated orthography, metaplasmuses, catachresis, enallage, and other methods galore, named and unnamed, for 'breaking' the rules.

What makes Mr Weller into a self-assured, metropolitan hero whose speech commands respect and also, vice versa, ascribes such speech to him is a fast-driving, stage-coaching system in particular—and a public transport revolution generally. This is why readers rightly credit this coachman with imaginative perceptiveness and not incompetence or insularity when he goes on in his letter to evoke his wife's death through a protracted coaching metaphor: 'her veels wos immedetly greased...'. Just as in Mr Weller's letter a speeding stage-coaching metaphor enters to obviate any reading of his letter as the homely product of provincial ignorance against the narrator's surrounding standard English, so too a historical transformation in passenger transport with coachmen like Mr Weller at its center creates a vernacular cosmopolitanism and helps buttress Dickens's elevation of everyday street language into print. Street vernacular, whose authority to describe the world is in danger of being subordinated to the narrator's standard English print, becomes empowered along with the historical transformation of the streets themselves.

In short, the language of the streets gets to sound modern in Dickens partly because of the modernization of the streets. Readers of Dickens have always understood that a 'primary cause of this author's success', as one reviewer ventured in 1837, is 'his felicity in working up the genuine mother-wit and unadulterated vernacular idioms of the lower classes of London'.[73] One might, justifiably enough, chalk up that success to the historic march of democracy in the era. Or, alternatively, as Bakhtin's work on carnival has perhaps most influentially suggested, one could assume that this resource was always there awaiting the ear and pen of a Dickens as it was at an earlier time for Rabelais, the degree of its availability and its tenor only alerting us to the vicissitudes of a wise folk-laughter always ready to clown around and upend the currently officiating officialdoms.[74]

At stake in all this was not merely that Dickens evokes the trammels of repetitious empty verbiage comparatively against the winged words of a vernacular street dialogue specially attuned to present realities, and that to do so he had first to hear the voices in the street as such.[75] The style—the battling discourses—cannot be separated from their broader referential content. Dickens is setting in melodramatic and moral opposition in *Pickwick* two different forms of systems, one epitomized in *Pickwick* by the law and the stasis of prisons, the other by the passenger transport network and the rolling road.

Long before Mr Weller's letter in Chapter 51, long even before Mr Weller's first appearance in Chapter 20, and right on the heels of the story's opening onto the bureaucratic buffoonery of the Pickwick club's minutes, the new accelerating street network to which Mr Weller's character belongs materializes to empower folk language not to sound haplessly folksy but to provide instead an equal and broad footing for challenges to bureaucratic institutionalizations of discourse and to those institutions giving rise to them. Once again recall Pickwick awakening on that May morning in 1827 in Goswell Street at the opening of Chapter 2, and how, self-importantly understanding himself to be embarking on intrepid travels though he will never get beyond the reach of a regional transit network, he packs up and heads for the nearest coach stand in St Martin's-le-Grand. There he cries, in Dickens's first novel's first quoted piece of dialogue: 'Cab!' This places him in the midst of an up-to-date accelerating transport network, even linguistically: in the late 1820s 'cab', short for the light two-wheeler cabriolet, was a new word. It marked the difference, which Dickens dramatizes in his sketch 'Hackney Coach Stands' (published in January 1835), between an older town taxi system that reused the gentility's cast-off hackney coaches and a new system based around the swifter two-wheeled cabriolets purpose-built for faster, short-distance, metered public transport.[76]

'Cab!' cries Pickwick, and in what at first might easily be mistaken for an incidental realistic topical detail indicating that the hero is on his way to his adventures, Pickwick calls forth the setting, temporally and spatially, of those adventures. In his first mishap, Pickwick, riding in the cab's open passenger compartment with the driver's seat attached beside it, questions his cab driver about his horse—how old is he? how long is he kept out at a time?—and copiously takes down the driver's answers in a notebook. The driver, literally and figuratively 'eyeing him askant' (1.5), naturally assumes Pickwick to be an informer, presumably from the SPCA, whose use of 'spies' to enforce Martin's Act (1822), the first animal cruelty law, was highly controversial.[77] So the cabman will coolly reply with outrageous lies: the horse is forty-two years old (actual average life expectancy was more like a dismal four); 'on account of his veakness' he stays in harness for two or three weeks at a time (a half day was more like the norm for a cab horse); and he does so because, as the driver concludes, neatly inverting the perceived implicit accusation of cruelty, 'he always falls down, when he's took out o' the cab' (1.6).

Dickens is staging here a primal scene of his work: the confrontation between everyday bustling street life and bureaucracy's record-

keeping attempts to control it. In this regard, the narrator firmly aligns
readers with the dialect-speaking driver and his deadpan linguistic
dodges. The extra laugh is, however, that the driver's dry irony is lost
because the encounter is not actually an instance of that conflict. The
narrator thus also keeps readers with Pickwick, whose gullibility nearly
beggars belief, and Pickwick's position begins to emerge fully for the
first time, carved meaningfully out of his not quite being the obnox-
ious notebook-wielding spy of legal authority that the driver takes
him to be. Instead Pickwick embarks here upon his heroic role—that
birthright commissioned by his name—of representative public transport passenger in relation to this cab driver, whose common name,
Sam, he even turns out to share. In truth, Pickwick (as passenger) needs
this cab driver's protective guidance, and thanks to a rapid authorial act
of characterological reincarnation and bald-faced renaming he will
subsequently get it from yet another equally sharp-tongued Sam
(Weller).

Dickens nonetheless can still make clear the general and central
nature of the conflict between bureaucracy and the streets that he has
evoked. When Pickwick arrives at the Golden Cross stage-coach inn,
where the three other Pickwickians are waiting to depart with Pickwick
for Rochester, the enraged cab driver refuses Pickwick's fare, states his
case to the people in the street—'Would any body believe as an
informer 'ud go about in a man's cab, not only takin' down his number,
but ev'ry word he says into the bargain' (1.6)—and having thus united
the 'crowd', including even those otherwise mortal enemies, hackney-
coach drivers and cabmen, proceeds to pummel the Pickwickians. In a
pattern that Dickens repeats in many of his novels, the second chapter
thus signals the answer that the story will elaborate in response to a
problem posed in the novel's first chapter, where that problem has
already been construed, knowingly and ironically, in anticipation of its
solution. Here the club members' capacity to club together and to
inscribe bureaucratically their own out-of-touch reality has come
ironically unhinged in the first chapter precisely at the moment Pickwick blusters that 'Travelling was in a troubled state, and the minds of
coachmen were unsettled. . . . Stage coaches were upsetting in all directions, horses were bolting, boats were overturning, and boilers were
bursting. (Cheers—a voice "No".)' (1.3). Then, in the second chapter,
the catching of a 'Cab!' to a stage coach and Pickwick's initial misad-
venture in doing so outline a reply to this first chapter's extended

display of self-aggrandizing bureaucratic absurdity ironically captured at the moment of dooming itself to a reconnaissance of the streets. (The story's initial two illustrations—the first depicting the Pickwick Club's board meeting at which the club's traveling branch is formed, and the second the street brawl with a cab driver—read together in just this oppositional way as a diptych.)

In its comedic figuration in *Pickwick*, the public transport system as a system subordinates itself to individuals' purposeful activity even as it, as a system, expansively merges individuals' circulating movement into a collective. Against this, Dickens juxtaposes the way other contemporary bureaucratic middle-class social institutions inversely threaten individuals. Where these other institutions fail is not in chaining people sheeplike to purposes that are not their own. Their failure, as Dickens relays it, follows from the purposelessness structured into systems, which otherwise crucially helps to give them their openness and publicness and allows their flexible extension into a future filled with purposes completely different from those of the present or past. As Dickens shows, this attribute returns in the corrupting form of individuals with no purpose other than to sustain the system, which in turn sustains them, but only as part of a system emptied of human purposes.

Such broken systems and the individuals in them look to be going nowhere, and a revolution in the passenger transport system, as Dickens renders it in *Pickwick*, especially makes them look that way. Consider Sergeant Snubbins, Pickwick's barrister, engirdled by his law papers, waving his pen, and suggesting to his flunky, Mr Phunky, that 'perhaps you will take Mr. Pickwick away'. This barrister dispensing with Pickwick is—as Dickens demolishes him—deeply absorbed in 'an interminable lawsuit' that concerns, like a microcosmic antithesis to *Pickwick*, 'the act of an individual, deceased a century or so ago, who had stopped up a pathway leading from some place which nobody ever came from, to some other place nobody ever went to' (11.327). Or, consider Mr Weller's comic pathos about greasing his dying wife's wheels to set her going. Suddenly one may hear it echo back across hundreds of pages as a rejoinder to the prosecuting lawyer's cynical invitation to the jury to grease Pickwick's wheels to make him pay.

Dickens did not identify the rot at the root of society around him and then, stumped as to alternatives, cry out wishfully for more and better morality, personal goodness, social 'humanity', as George Orwell claimed and so many others have believed prior to him and since (from Forster to Flint).[78] Dickens did not, that is, counter the 'systematic villany' (12.358) of some of his contemporary bureaucratic institutions merely with a kind of hazy-dazy, male, secular, bourgeois, loose reformist humanism in thrall to a falsely apolitical idyll of gendered, middle-class domesticity, while also remaining alert, with an eye sharper than perhaps had ever before existed, to the irreducible particularities of individuals and the endless strange quotidian minutiae comprising everyday human life. All together that has been taken as indicative of Dickens being essentially anti-system—'*against* institutions', as Edmund Wilson declared.[79] There are many censures that can be leveled at Dickens, but ultimately this one and Orwell's—'he has no constructive suggestions'—is unfair.

In Orwell's case, the impeachment stems from trying to look at Dickens politically. And the problem is not that Dickens's politics are incoherent and variable across his work and life (whose aren't?), nor even that in considering specific biographical instances (say, his work reclaiming prostitutes at Urania house), it is slippery framing Dickens politically—is he a bastion of middle-class morality? Or, as he edifyingly reappeared in 2007, a radical?[80] The problem is that trying to discover what Dickens is standing for politically, however broadly one might conceive of politics, is, in this instance, barking up the wrong tree. Politics? 'I object to the introduction o' politics.... it ain't true. I say that coachman did *not* run away; but that he died game...' (15.464). The public transport system is no political system, to assert what Sam Weller would probably deride as a 'self-evident proposition'. Though one can look at the politics of transport, as one can look to the politics of almost anything, in writing about the public transport system Dickens was not offering a political alternative—neither an economic one, nor a religious one, nor, despite all the final wedding bells, a gendered domestic one—which is not to deny for a moment that his novels are crammed full with politics, class, religion, gender, and so on, or to suggest refraining one iota from looking at his novels in all such terms.

Dickens has always been justifiably famous for his brilliant parodic anatomizations of the individual in modern bureaucratized society. If anything, this regard has only expanded in the aftermath of the influential work of Michel Foucault and D. A. Miller, who illuminated the novels' own potential participation in constructing self-surveying, discursively formed subjects caught in the middle-class's newly diffuse disciplinary institutions. Yet, amazingly given Dickens's unmistakable, razor-sharp, comic capacity to look inside and across such systems, it has also been relatively commonplace to believe both that Dickens somehow ingenuously pits individuals and individuality against those soul-crushing administrative systems *and* that, alack, Dickens is perhaps also not the best place to go to follow the interior psychological development of individual character.

Instead—I have suggested—one might recognize not only that Dickens sharply distinguishes the passenger transport system from its close confederate, the postal system, along with its literary corollary, the epistolary novel, but also that he makes that public transport system into something like the grounds, albeit shaky, from which to launch his comic critique of deadening bureaucratic systems along with their attendant tongues (those obfuscating jargons and turgid circumlocutions that Dickens delights in speaking tongue-in-cheek). As we will see, Dickens would increasingly recognize the impossibility of sustaining such an opposition. He did not, however, therefore put it behind him. Rather, he explored its breakdown—in a tragic vein in *Master Humphrey's Clock* and later, definitively, in *Little Dorrit*—as his view of the public transport system evolved.

2

On Tragedy's Tracks

I. In *Clock*

The inland public transport revolution, as Dickens envisioned it in *Pickwick*, ideally aimed to expand ordinary people's individual mobility and even comprehensively to encompass all in its network. In reality, however, it also transformed walking on the intercity highways into a comparatively slow, onerous, and solitary mode of journeying and re-created it as the possibility of falling out of its network. This redefinition in relations of mobility was not quite so stark or real as long as the shuttling stage-coach passengers and wayfaring pedestrians were at least sharing the road. The railways changed that. They solidified the divide, imposing a separate and separating physical structure of bridges and tunnels, trains and tracks, stations and high-speed corridors.[1]

The railways lifted, as it were, the newly busy long-distance road traffic and laid it down separately in a system running right next to those roads. The stage-coach inns—where horses were changed—died away, and Wilkie Collins issued an essay-epitaph in 1843 for 'The Last Stage Coachman'.[2] All around the railways, horse-drawn road traffic continued to grow, but the railways otherwise emptied the intercity direct roads of their long-distance travelers.

Besides those who chose to hike for pleasure (or other reasons) and those whose livelihood kept them on the roads, left behind was an indigent minority forced to make their long-distance journeys by foot. This debarment of the poor counts as one among many modes that the public transport system has historically created for expressing social inequities even as it has also offered challenges to their reproduction. As we all know, segregation and differential accessibility form an essential part of passenger transport history. Rather than chart how public transport has translated or helped to create inequitably discriminatory relations defined by economics or gender or race or empire or other such potent stratifications of human social relations, my concern in this book is with public transport's creation of an autonomous site of communal relations where such divisions of community may be enacted. For this reason, it may seem that I overstate the network's inclusivity and understate how, like any network, it is also about exclusion. Yet theorizing public transport's networking can help, I believe, to illuminate its use for divisive discrimination—in, say, to evoke one notorious instance, the racial segregation of railways.[3]

Accordingly, this chapter takes up some of the passenger network's own self-created limits and intrinsic failures in constructing its networked community. And in this regard, one can see that the railways not only had an exiling effect on intercity foot-travelers that revealed the possibility of dropping out of the network altogether, as I mentioned above. The railways also, conversely, foregrounded how their own passengers were sacrificing the kind of independent, individual control generally understood to inhere in walking and instead magnified how their passengers' coordinated journeys inexorably bound them to others in the network.

Dickens fashioned, I am going to argue, precisely these relations of mobility into a tragic novel. The novel in question fully qualifies as an Aristotelian tragedy because it shows that the same forces holding together a community ironically rip it apart, and it thus lays bare the community's inherently flawed form.[4] In this novel, Dickens first explicitly figures *Pickwick*'s extension into the railway era, reviving the Pickwickians. In a tale then told to Pickwick, Dickens returns to the stage-coaching days of *Pickwick*. This time, however, Dickens dramatizes the story of a foot-traveler and her companion who leave London and proceed on a one-way trek to the industrial north, disappearing out of the reach of their pursuing friends and enemies, until finally, fatally debilitated by her exhausting trudge, the heroine sinks into her last resting place.

This tale was the best-seller that appeared under the title *The Old Curiosity Shop* in the pages of *Master Humphrey's Clock*, and little Nell is the famous child heroine whose fatal foot-journey it relates. In this chapter I am going to examine this tale, first titled *The Personal Adventures of Master Humphrey: The Old Curiosity Shop*, in a way that it has never been looked at critically before: as a tale told by Master Humphrey to Pickwick in a railway era. That is how it appeared originally in its serialized form, how its first readers read it, and there, in *Clock*, is where the tragedy containing the tale *The Old Curiosity Shop* both begins and ends.[5]

In *Master Humphrey's Clock*, Dickens revived Pickwick and the Wellers. Steven Marcus is thus not hypothesizing some analogical correspondence when he suggests that 'in *Master Humphrey's Clock* [Dickens] made an effort to re-create the circumstances of *Pickwick Papers*', that 'Master Humphrey's club is a bizarre distortion of the Pickwickians', and that Master Humphrey's 'situation and the quality of his response to it are simply the opposite of Pickwick's'.[6] As Master Humphrey himself declares, 'Mr. Pickwick and I must have been a good contrast.'[7] The important question here is, however, just how and why this return of Pickwick and the Wellers 're-creates' as a 'bizarre distortion of the Pickwickians' *Clock*'s imaginative framework—a late-night reading club in which four old men draw stories from an old pendulum clock case filled with manuscripts they have written, their leader, Master Humphrey, a reclusive, nocturnal *flâneur* as well as *Clock*'s narrator.

This crucial relation needs recovering partly because, notwithstanding Marcus, it has become a widespread mistake in discussing *The Old Curiosity Shop* to write off the *Clock* materials with the half-truth that Dickens saw declining sales and swooped in to rescue the magazine with the novel. Not least this tidy stopgap economic explanation cannot survive the complicating fact that, having expanded the upstairs Clock reading club to a downstairs 'Watch' coterie with the Wellers, Dickens fully keeps his promise to tell of the doings of 'Mr. Weller's Watch' four chapters into *The Old Curiosity Shop*. Reading *Shop* in its original form in *Clock*, one thus reads of the Pickwickians after the narratorial baton passes at the end of the third chapter of little Nell's tale from Master Humphrey's first-person recounting to an omniscient narrator (a shift that definitively indicates to readers that the tale had metamorphosed into a novel). And, even further on, after the tale's eighth chapter, Dickens dilates further upon the doings of Mr Weller (3.128–32). (For those counting, this last time Dickens interlards the Pickwickians comes in weekly number eleven, midway through the third monthly number of the total eleven monthly numbers in which the tale *The Old Curiosity Shop* ran.) Nor does it take a Steven Marcus to discern the contrapuntal relation between little Nell's building woes

with her grandfather and the extended picture Dickens interweaves of
Mr Weller as a proud grandfather joyfully parading his grandson around
in a mini-coachman's get-up—'ya—hip!' (3.130), such 'a angel!' (3.131),
'bless your heart you might trust that 'ere boy vith a steam engine
a'most' (3.132). Not so very curiously, this contrastingly comedic
grandfather-and-child portrait concludes happily (forebodingly for
the little grandchild in *Curiosity Shop*) with Mr Weller 'carr[ying] the
child like some rare and astonishing curiosity' to perform comedy
routines 'with the utmost effect to applauding and delighted audiences'
(3.132).

With regard to this vital, overarching relation between Master
Humphrey's clock-club with its Pickwickians members and the
tale that Master Humphrey tells them, in which a very different
protagonist and her sidekick, two 'adventurers', leave London by a
very different mode of transport also to wander they know not
whither, one might indulge a trivial, sadistic, Quilplike pleasure—
'Here's sport!...sport ready to my hand, all invented and arranged,
and only to be enjoyed' (5.211)—in imagining springing the frontispiece for *Clock*'s first volume upon all those modern unwitting
readers of its abridgement into the novel *The Old Curiosity Shop*.
Those readers would be completely dumbfounded to discover
George Cattermole's illustration erupting fully halfway through their
story, at Chapter 37 (Fig. 10).

Some recognizable characters from *The Old Curiosity Shop* appear
along the margins of Cattermole's illustration, set amid the decorative
angels and knights. This being the frontispiece to *Master Humphrey's
Clock*, however, the artist pictures front and center an upstairs with
Master Humphrey across from Pickwick asleep in front of the clock
and a downstairs with the Wellers and company, all snoozing as well.
Coming so long after the tale *The Old Curiosity Shop* has completely
taken over the pages of *Clock*, the picture's discombobulating effect of
just what frames what could not be more perfect. Indeed, as Chesterton saw, Nell's story is significant and powerful partly precisely because
Dickens shows how an incidental encounter with a poor child on a
London street can expand into an entire novel about the brief life of
that seemingly insignificant and marginal individual, which narrative
expansion, in return, gives meaning to her story's compressed origins
in a short tale.[8] For more reasons than one, then, this tale belongs to
the margins, as Cattermole draws it.

Figure 10. The Frontispiece to Volume One of *Master Humphrey's Clock*. Courtesy of the Department of Special Collections, Young Research Library, UCLA.

The frontispiece holds equally true to *Clock* and its re-engagement of Pickwick and the Wellers. And it is not merely that Pickwick almost always did fall asleep whenever tales were told in *Pickwick*. The mythic medieval past woven fantastically and artifactually around the central present-moment (itself fittingly divided into two simultaneous scenes) cues readers to the picture's framings in historical layers. The image thereby figures how *Clock* relates to *Shop*: where Master Humphrey's clock-based reading club transforms the Pickwickians from the adventurous protagonists of a fast-moving, stage-coaching, traveling club into the slumbering members of someone else's reading group, Nell's saga relates a peripheral history that only emerges after *Pickwick* retires.

Gone from the Pickwickians, in a palimpsestic sort of way, is the stage-coaching apparatus that gives its name to *Pickwick* and makes for its—and his—going. One inspired contemporary reviewer (Thomas Hood) beautifully catechized Dickens for just this seemingly absurdly explicit lack in the voice of the coachman Mr Weller, demanding 'what literary new, fast post-coach could make a more hock'erder start than with four insides, professedly booked to nowhere at all, and with such a wery unconwenient time-keeper as an old, wenerable, antiquated eight-day clock on the roof of the vehicle?'[9] 'Vy, nonesomever', one may readily concur, and yet, not being reviewers merely gladdened to see the eclectic *Clock* materials' gradual displacement by the serial novel *The Old Curiosity Shop*, also wish to think through the two together. After all, as one might quiz Hood, the tale's slow-going, booked-for-nowheres protagonists whom Hood admires so much supposedly by contrast hardly represent a return to the fast-coaching adventures of *Pickwick*. In fact when Hood rejoices that 'the author quietly gives [the Pickwickians] the slip and drives off to take up characters, who really have business down the road', the whole point would seem to be that *Clock* sets up the journey of little Nell in precisely such terms, as a giving of the slip, as it were, to the Pickwickians.

What this means is not only that *Clock* casts Nell's foot-journey as occurring outside of the public network for journeying whose rise *Pickwick*, both the novel and the stage-coaching line that gave it its name, signified. That's true, but also, where in *Pickwick*, in 1836–7, Dickens rendered a story of regional, late-1820s stage coaching silently shaped by the present-day onset of the railways, now four years later in *Clock* Dickens pointedly shifts his readers into that railway present. That consummate stage-coachman, Mr Weller, is now outdated, grandfathered even—'it wos all wery shockin'' (3.132). 'Outrage and a insult': he is, like other coachmen and guards, 'forced to go by [rail]' as a passenger. 'I wos a goin' down to Birmingham by the rail, and I wos locked up in a close carriage with a living widder', Mr Weller recalls, denouncing the practice of locking rail carriages from the outside (which exacerbated fears of sexual assault and fears of false claims of it). Then follows his longest speech: a comic monologue about the railways as seen from the viewpoint of a displaced coachman (2.72).[10]

In this semi-serious diatribe about the railways, the old coachman signals the tragic vulnerability of the networked community that Nell's foot-journey will subsequently bring, as the phrase goes, to its knees. Mr Weller's opening sally, also topic sentence, strikes the keynote: ' "I con-sider" said Mr. Weller, "that the rail is unconstitootional and an inwaser o' priwileges, and I should wery much like to know what that 'ere old Carter as once stood up for our liberties and wun 'em too—I should like to know wot he vould say if he wos alive now, to Englishmen being locked up with widders, or with anybody, again their wills. Wot a old Carter would have said, a old Coachman may say, and I as-sert that in that pint o' view alone, the rail is an inwaser" ' (2.72). Homonymically recasting the legendary originary constitutional formulation of communal English individual rights against monarchial prerogatives, a.k.a. the Magna Carta, as the heroic act of some archaic cart driver—an 'old Carter'—who is his progenitor, Mr Weller jokes wittily but truly here that the revolution in public transport, which he first embodied and whose continuation has now displaced him, concerns nothing less than the passenger transport system's capacity to constitute community by merging mobile individuals at liberty to pursue their own paths.

This is familiar. The collectivity created by the merging of individual journeying that Mr Weller invokes and that partly defines the rise of the public transport system has virtually also been a repeating topic sentence thus far in this book. And it would be possible to rattle on here about how Mr Weller's speech can be seen to follow and flow from 'Mr. Pickwick's Tale', the longest intercalated tale in *Clock* printed before the tale *The Old Curiosity Shop* and the only one interleaved with it. 'Mr. Pickwick's Tale' looks back to the seventeenth century to a cart-driving hero who demonstrates transport's primordial power to forge community across an unpaved, unconnected world, riven by religious turmoil over Catholicism and by superstition so deep that there are witch traps set on the public highways.

But more germane at this point is how, according to Mr Weller, the railways threateningly amplify the merging of passengers' individual journeys. As Mr Weller sharply asserts, this poses a real problem for the historical restructuring of society around its expansion of individual freedom of movement. Nor is the problem delimited to the inside of the carriage compartment. In his next line, Mr Weller looks outside the carriage and reports on a troubling, mirroring, de-individuating regularization of the trips. Protesting against the amplification of the standardization of space that I discussed back in Chapter 1, Weller disapprovingly catalogues at mimetic length the rail journey's iterative details of 'never comin' to a public house, never seein' a glass o' ale, never goin' through a pike, never meetin' a change o' no kind (horses or otherwise), but alvays comin' to a place, ven you come to one at all, the wery picter o' the last, vith the same p'leesemen standing about, the same blessed old bell a ringin', the same unfort'nate people standing behind the bars, a waitin' to be let in; and everythin' the same except the name, vich is wrote up in the same sized letters as the last name and vith the same colors' (2.72).

Readers are supposed to see here, as throughout his speech, right through Mr Weller's disgruntled insider perspective. After all, the rails, upon which he self-confusedly aspires for his grandson to become an engineer, continue the passenger transport revolution that made him before displacing him. The recurring scenes of standardized railway travel that he lists thus only intensify the recurring scenes of standardized stage-coach travel. One need not have ever ridden a stage coach to suspect that authentic 'change' hardly derives from repeatedly 'changing horses'.

In other words, Mr Weller does not quite slide out from under his own critique, and that critique questions wholesale the public transport revolution's equation with individual freedom of movement. The railway passengers, locked up in their carriages, their trips regimented, find themselves stuck 'behind the bars' at the policed station. Offhand, one could pretty easily, and essentially correctly, follow such carceral metaphors into an interpretation of the railways in terms of the historical, panoptic-disciplinary administrative reorganization of society, synonymous with Foucault. Following that familiar vein into its discursive diffusion, one would notice, for instance, how power here infiltrates even into the standardizing of print, such that readers may find themselves reading—self-observantly—about 'the same sized letters'. Remaining focused instead on the passengers' journeying, the basic point is simply that their individual mobility has become their collective entrapment. To cross-channel the French theorist and historian of speed, Paul Virilio: 'The events of 1789 claimed to be a revolt against subjection, that is, against the *constraints to immobility*...a revolt against arbitrary confinement and the obligation to reside in one place. But no one yet suspected that the "conquest of the freedom to come and go"...could, by a sleight of hand, become an obligation to mobility', a 'dictatorship of movement'.[11]

Clearly Dickens's earlier opposition between the Fleet prison and Pickwick's speeding coach is in trouble. Later, when he paints the prison-world of *Little Dorrit* (also set in the 1820s), he will craft a perspective capable of expressing their dual relation. At this point, in Mr Weller's speech, what becomes evident is that the railways expose, partly by further consolidating, the darker underside of the truism that merging passengers' journeys merges their journeys.

In Mr Weller's speech's grand finale, he especially suggests how this passenger transport history stipulates Nell's tragedy. Evoking how the railway's acceleration makes individual journeys indivisible, Mr Weller proclaims the indefatigable train constitutes a tragedy in the making: 'as to the ingein as is alvays a pourin' out red hot coals at night, and black smoke in the day, the sensiblest thing it does in my opinion is, ven there's somethin' in the vay and it sets up that 'ere frightful scream vich seems to say "Now here's two hundred and forty passengers in the wery greatest extremity o' danger, and here's their two hundred and forty screams in vun!"' (2.72).

Evaluating this description spatially, the train passengers screaming along call attention to the railways' historic construction of purpose-built high-speed passenger transport corridors. Prior to the railroads, collisions on the road—'there's something in the vay'—essentially occurred between individuals, albeit often grouped, who traverse a shared route. Such collisions even signified a failure to negotiate that route as shared. By contrast, the very indistinguishability registered by 'there's something in the vay' not only indicates the railway's accelerated speed's eye-blurring of that 'something', but also the undifferentiating othering the train creates by necessarily proscribing anything but its kind on its road. Pausing presciently to think of Nell's trek, one could extrapolate a warning that the railways cut off passengers from highway pedestrians, who are no longer even moving spatially along the intercity networked routes. But, even more directly, Mr Weller is registering the other side of that exile: how those on the railways find themselves rolled together.[12]

And, in this regard, Mr Weller's dark view of an annihilating space comes along with a related one concerning time. His description that the potential impending rail calamity rolls together 'two hundred and forty screams in vun' is his creative interpretation of the steam-whistle, but it also registers how, while ordinarily passengers flow through the public transport network bent upon their individual pursuits, something either in imagination or in reality threatening the speeding journey can suddenly make real their merging of those separate journeys, actualizing the passengers' unification. Most extremely, because those passengers temporarily co-locate and combine their individual journeys, they also accede to the possibility of a shared fatal ending.

This present unity that coalesces for public transport passengers around imagining an impending crash indicates how, for those inside the carriages, the acceleration into which they blur their individual journeys helps meld, at another level, their separate diachronic plot trajectories into a collective simultaneous time. The crucial thing here is that each individual passenger's time—time made from their joint motion across space—becomes one time, simultaneous and shared, not quite, as it might perhaps seem to those inside the carriage or to someone thinking of them, solely by virtue of their simply coming together in the same place. What place? They are flying between places. From the perspective of their pocket watches, Weller's rail passengers are charging along only too fast. In fact, their rapid merged movement by a passenger transport system is reconstituting their time as synchronized across long distances. This was the mind-bending time warp that was emerging in the 1840s, not directly articulated in Mr Weller's speech, not yet perhaps in fact historically quite fully articulable, but following from his insight that people's rapid circulation newly interlocked them together. (Its relevance becomes manifest at the end of the novel when Master Humphrey's clock retakes center stage.)

In the 1840s, for passengers inside the mobile rail carriages the time being carried from the public transport system's London hub manifestly began to face off against all the regional town clocks declaring their different local times and even to face off against the sun, that heavenly chariot, which had been telling everyone below that it was noon only when it approached its zenith. Those local times and longitudinally discrete sun-readings were useless and confusing to the people inside the trains in rapid motion. Why should the watch of the passenger who just got on at Bath be out of synch with that of the passenger coming from London?[13]

And that is just the half of it, as retracing the inside-outside logic of Weller's speech helps make evident. For having shown first that *inside* the rail carriage there is an invasive merging of journeys and then *outside* a corollary standardizing of trips, our disgruntled stage-coachman, defeated by the railway's speed, further observes that the dangerously unstoppable railway speed *inside* the carriage binds everyone together and...and there his speech ends. Following this inside-outside composition through to its final missing outside conjures the question: What is that railway speed doing then outside those speeding trains at the stations and along the route? The answer, which coheres with what is going on inside the carriage, is that the railway's speed is systematizing time into one for the people outside at the stations and in all the towns it networks. If you left London by train an hour ago, why shouldn't it be an hour later in the town where you arrive?

This standardized time, with its contemporary contemporizing properties, first came up in discussing *Pickwick*. There it was noted that the swift Royal Mail coaches (like the one that Nell will see 'dashing past like a highway comet, with gleaming lamps and rattling hoofs', 8.54) clove to an onboard clock set to London time (often regulated to gain time for eastward travel and lose for westward).[14] During the stage-coaching era, this network time was as yet invisible to the public. Essentially this clock marked the stage-coaching system's internally synchronized space-and-time self-discipline, part of its tightening management of when and where. Nonetheless, from the perspective of the mail-coach men speeding London time around and about the nation—from a perspective, that is, embedded in the accelerating regional public transport system—a reconventioning of time had begun through the ever-swifter networked circulation of people and the coordinating of their meetings and partings. Hence in 1840 the Commissioner for Longitude could write to the Postmaster 'propos[ing] to regulate all post-office clocks in the Kingdom, by means of the time brought from London daily by the mail-coach chronometers'. He had 'no doubt that, ere long, all the town clocks, and, eventually, all the clocks and watches of private persons, would fall into the same course of regulation; so that only one expression of time would prevail over the country'.[15]

The intercity inland public transport system, accelerating its journeys via the trains, could not wait on the postmaster's decision. As Weller's speech indicates, in the early 1840s a national railway network was coming together. The railways had been overtaking and replacing the long-distance coaching system for a decade. Standardizing these speeding trains' timetabling was not, as it is sometimes thought, absolutely required to avoid fatal crashes. Dangerous proximities could be indicated by mechanical semaphores or human 'signalmen' (later the

titular subject of one of Dickens's best short stories). The trains' schedules did, however, need to be coordinated with unprecedented precision across unprecedented distances. Plotting and tracking their rapid
trajectories necessitated imagining activity—a departure, an arrival—
occurring at towns many miles apart in concurrent time: clocks ticking together across a terrestrial grid.

In November of 1840, just a few months before the February 1841
close of *Shop*'s serial run in *Clock*, the Great Western Railway mandated the London time kept by St Paul's for its schedules. The railways
thus began to make patent and public the clock recalibration that the
public transport system had been building. Many town clocks began to
display two separate minute hands. One hand indicated local time, the
other railway. Time itself was, for a time, out of joint. The new 'railway
time' was first based on London time kept by St Paul's and later
switched to Greenwich mean time, or GMT (differing by 23 seconds
from St Paul's London time).

The new standardized time offered some practical benefits to the
public. It improved people's coordination of their activities with each
other as well as with the public transport system. It began the process
of doing away with the annoyance, as one railway schedule bewilderingly explained, that 'London time is...four minutes earlier than
READING time; 7½ minutes before CIRENCESTER; 11 minutes before
BATH and BRISTOL time; and 18 minutes before EXETER time'.[16]

More profoundly, while the change only amounted to a matter of
adding or subtracting a few minutes from clocks, standardized time
recalibrated the local times based on communities unified by their
geographically proximate terrestrial place under the sun. That ancient
configuration of a temporally-bonded community fused with another
formation whose shared time and space was established by people's
continual, fast circulation and interconnectedness by an intercity public transport network. But where, in *Pickwick*, that accelerating network had heralded a newly shared space and time that united individuals
simultaneously pursuing their separate paths, in *Clock* Weller's critique
of the railway system's excessive binding together of passengers' journeys signals a snag.

The problem is that when the rapidity of the trains knit passengers
inside and outside together in a synchronized time, enabling all to better clock their crisscrossing lives in its shared arena of activity, that
mobile networked community's synchronized time also came together

in a totalizing form inexorably mutually attaching. Insofar as standard time sprang from the passenger transport system (bracketing, for instance, its important relation to industrialized capitalism), it arose as—and still is—the property of a networked community to which individuals were subordinated not because it dictated to them in particular ways how to use their time, but because it helped define the manifold ways they did as a systematic relation of synchronized, circulating coordination with everyone else. In clocking themselves together, fellow travelers actually reinforce with each individual journey the totalizing temporal system of their collective circulation. And yet because this comprehensive collectivizing rests upon a physically limited passenger transport network, it also follows that it became possible to walk right out of it.

II. A Tale That Is Tolled

Once Nell leaves London, her long journey reflects the assorted hardships inflicted upon her as a homeless, motherless, young girl threatened with sexual exploitation and saddled with the care of her senile, gambling-addicted grandfather. When she leaves the city, Nell both flees from her many difficulties and carries them with her: walking the streets to escape walking the streets, hitting the road to try to settle down, sleeping at a factory rather than sleeplessly 'work, work, work' in one, rambling on rather than gambling on—'oh, do let me persuade you . . . to try no fortune but the fortune we pursue together' (6.269).

But, as a tale told to the Pickwickians in *Master Humphrey's Clock*, Nell's fatal foot-journey also signifies in terms of a wider history of railways and long-distance stage coaching. In the context of the contemporary passenger transport revolution, Nell's walking forth on a long-distance journey severs her from her community. (That is her goal: to escape.) As Chesterton has remained virtually alone in noticing, Dickens's 'artistic idea' in this novel was not actually to showcase Nell's death. His artistic idea was 'that all the good powers and personalities in the story should set out in pursuit of one insignificant child, to repair an injustice to her, should track her from town to town over England with all the resources of wealth, intelligence, and travel, and should all—arrive too late'.[17]

As a tale of that chase, *The Old Curiosity Shop* reads as the spoiler to *Pickwick*. Not merely its antithesis or its repudiation, *Shop* represents a tragedy derivable from *Pickwick* that further illuminates its networked relations. And when looked at thus, not only Nell and the grandfather's intercity walk and the racing coaches arriving too late pop out, but also, and inseparably, the novel's main structuring split into two extended, alternating, retrospective narratives. Carpe diem meets corpus delecti; 'Where shall we go next?'—'Where is she?' Alternating Nell's trek against the unfolding plots of the London-based characters, who wish to include her and fail, twice chasing after her by post coach, Dickens represents Nell's falling out of the public transport system as a falling out of communal time and space, and he brings to light how the novel's multiplotted form works to construct that shared time and space.

To begin with the tale's beginning, once one recognizes that as a tale introduced to the Pickwickians of *Clock* Nell's foot-journey signifies against a public transport system she does not access—placing it in a bigger history of railways and long-distance stage coaching—one can also suddenly discern how internally *Shop* pivots structurally on the difference between walking in London, where it is a common means of mobility, and walking upon the intercity highways—those 'main roads [that] stretch a long, long way' (5.236)—where pedestrianism has come to signify exclusion from a network of public mobility. Over the course of the first twelve chapters, a pedestrian encounter in London opens a tale that Dickens expands into a multiplotted novel in which Nell and her grandfather hold a primary place. And then, having built up this multiplotted novel by tracking its ensemble of characters here and there about London, Dickens has Nell and her grandfather escape by literally walking out of it: 'Let us steal away . . . and leave no trace or track for them to follow by' (3.154).

The tale opens: 'Night is generally my time for walking' (1.37). Boz-like, Master Humphrey, as the unnamed, first-person narrator, introduces himself as a nocturnal *flâneur* 'speculating on the characters . . . of those who fill the streets' (1.37). The very first thing this narrator Master Humphrey wants to get across, however, is the oppressive inescapability in the city of 'that constant pacing to and fro, that never-ending restlessness, that incessant tread of feet wearing the rough stones smooth and glossy' (1.38). Master Humphrey: 'Think of a sick man in such a place as Saint Martin's court, listening to the footsteps, and in

the midst of pain and weariness obliged, despite himself...to detect
the child's step from the man's, the slipshod beggar from the booted
exquisite, the lounging from the busy, the dull heel of the sauntering
outcast from the quick tread of the expectant pleasure-seeker...think
of...the stream of life that will not stop, pouring on, on, on' (1.38).

The follow-on to this opening onto the crowded, coursing city
streets, where, each atomized from the other, all walks of life process
together, is its enactment. In those urban streets filled with foot-
passengers, the well-off old narrator randomly crosses paths with a
young girl: 'But my present purpose is not to expatiate upon my walks.
An adventure which I am about to relate...arose out of one of these
rambles.... One night I had roamed into the city, and was walking
slowly on in my usual way...when I was arrested by an inquiry...and
found at my elbow a pretty little girl, who begged to be directed to a
certain street at a considerable distance...(1.38). It would seem only
too easy to link Master Humphrey's description of how they 'trudged
away together: the little creature accommodating her pace to [his], and
rather seeming to lead and take care of [him] than [he] to be protect-
ing her' (1.39) to Nell's subsequent long-distance journey, in which she
similarly guides her grandfather. The difference between walking in a
city and walking between cities, however, makes all the difference.
Becoming lost in London, Nell is quickly reoriented and returned
home. And though she cannot afford to hop into a hackney coach to
get there as Master Humphrey can, still London's pedestrian streets
mean their separate paths may spontaneously entwine. There is for
readers as yet no problem believing Kit when he asserts (in words that
later become haunting): 'I'd have found her master...I'd have found
her. I'd bet that I'd find her if she was above ground, I would, as quick
as anybody, master' (1.43).

Lose Nell in London? Again: over the course of the first twelve
chapters Dickens makes it increasingly apparent that she is surrounded
by people with their sights firmly set on her. This is made true not
only in a literal sense, as when Master Humphrey and Kit both secretly
attentively watch her from the street as she gazes down at it from her
attic window, but also—and more intensely—plotwise. Dickens develops virtually all the various characters such that in one way or another
they hold Nell, along with her grandfather, in the cross hairs of their
joint future. It is not necessary to review here in any great detail how
Nell's rogue brother Fred Trent hatches a mercenary scheme for his

accomplice Dick Swiveller to marry Nell that forces Swiveller to
abandon a romance that 'man never loved that hadn't wooden legs...
[his] heart... breaking for the love of Sophy Cheggs' (8.76), or how
Quilp, proposing darkly that Nell replace his Mrs and pre-emptively
taking up sleeping in Nell's bed, sets up shop in *Shop* as Nell's 'evil
genius' (10.185). Part of Quilp's Richard-the-Third-like function on
the one hand and Swiveller's as the self-aware pawn on the other is to
make clear that while readers are not likely to dismiss the contents of
the plotting, still a key aspect is less those contents than the all-
enmeshing fact of it.

In just this sense, readers encounter Nell and her grandfather's secret
departure as a plot event whose meaning lies partly in the deflating
threat it poses to the very continuation of the building, multiplotted,
serial novel. Here is just one example of several that read virtually
metafictionally: 'Richard Swiveller was utterly aghast at this unexpected alteration of circumstances, which threatened the complete
overthrow of the project in which he bore so conspicuous a part, and
seemed to nip his prospects in the bud.... [H]e had come... prepared
with the first instalment of that long train of fascinations which was to
fire [Nell's] heart at last. And here, when he had been thinking of all
kinds of... approaches... here were Nell, the old man, and all the
money gone, melted away, decamped he knew not whither, as if with
fore-knowledge of the scheme and a resolution to defeat it in the very
outset, before a step was taken' (4.160). Quilp is equally and similarly
vexed, so too Kit.

Yet, as every reader knows, in the very first outcry over Nell and her
grandfather's absence, the London-based characters' plots, far from collapsing, thicken and more deeply intertwine. Immediately after Nell
and her grandfather leave, the two main, hitherto separate, threads of
intrigue—embodied in Swiveller and Quilp—literally tangle with
each other at the door left unlocked between them. Shortly thereafter,
Quilp will revive Fred Trent's mercenary plan for Swiveller to marry
Nell because it offers Quilp an irresistible buffet of vengeance upon
everybody. It's quite a reversal. The scheming goes on apace 'around'
Nell whether the loot is there or not, with or without her around. By
the time that, unbeknownst to them all, Nell collapses on the road and
is carried to the church in which she will be buried, the plot has
become so complexified in London that even the most careful readers
are likely to lose sight of the fact that Quilp frames Kit for theft—'I shall

have to dispose of him, I fear' (8.65)—because he wants Kit out of the confidence of the single gentleman, whose blackmailing about the grandfather's madness Quilp is now improbably scheming upon: 'But for these canting hypocrites, the lad [Kit] and his mother, I could get this fiery [single] gentleman as comfortably into my net...' (8.65). But readers needn't worry overmuch. The thing to grasp is that after Nell's exit the London-based half of the novel continues to develop in the multiplotted form first linked to 'the crowds for ever passing and repassing' on the streets of London (1.38).

Minus of course—crucially, increasingly desperately, and ultimately tragically—Nell and her grandfather. Directly upon their departure a chorus begins, its refrain recurring at regular intervals through to the tale's end. Its repeated plaint is 'Where is she?':

'Come here you Sir,' said the dwarf. 'Well, so your old master and young mistress have gone.'

'Where?' rejoined Kit, looking around.

'Do you mean to say you don't know where?' answered Quilp sharply. 'Where have they gone, eh?'

'I don't know,' said Kit. (4.162)

'Forth from the city, while it yet slumbered, went the two poor adventurers, wandering they knew not whither' (3.156): so ends Chapter 12, weekly number thirteen, and the third monthly number of *Master Humphrey's Clock*, and from that carefully structured break on, the novel dramatically divides itself between following Nell sequentially on her foot-journey and the multiplotted realm of the London-headquartered characters. (See Fig. 11.) Back and forth goes the story, switch points always tabulating with chapter breaks: 13, 15, 20, 24, 33, 42, 47, 52, 56.

This is as serious a formal experiment as that of the more famously split narrative of *Bleak House*. And though the alternation between Nell's journey and the action based in London does not shift between an omniscient, third-person, present-tense narrator and a first-person, past-tense account, Dickens almost as dramatically cleaves this story between two competing narrative forms.

Initially the split appears largely a matter of focalization. The omniscient narrator jumps between characters in London, providing a strong sense of their ongoing simultaneous activity and their converging intersections. By contrast, the narrator accompanies Nell on her

Figure 11. Nell and her grandfather outside London. They have left for (arbo)real. The building most visible in London is St Paul's. Courtesy of the Department of Special Collections, Young Research Library, UCLA.

journey sequentially, holding closely to her viewpoint and to narratorial reflections on her immediate and local surroundings. Nell's narrator never leaps remotely away from her to describe what's going on in her absence—to tell us about, say, the separate doings of her grandfather or Mrs Jarley. The description of Codlin waiting for Nell's arrival at the Jolly Sandboys Inn is the self-conscious exception to the rule (the action blatantly on pause till she arrives), and the way in which the reader discovers unexpectedly along with Nell that Mrs Jarley has seen her earlier at the races or that the man ahead walking is the schoolmaster indicates how such crossings of paths unfold in her portion of the narrative focalized through a limited third-person point of view.

Nell's journey thus initiates a division between a single-focused narrative plot thread and the multistranded plot activity in London. What's complicated about this split is that the multiplottedness of the ensemble in London itself derives from alternations in which the narrator follows a single character's plot, and then another's, and so on. As Peter Garrett explains: 'the form of these [Victorian multiplot] novels is neither single- nor multiple-focus but incorporates both, and it is the interaction and tension between these structural principles which produces some of their most important and distinctive effects.' The switches to Nell on her journey maximally exploit (to quote Garrett again) 'the centripetal impulse that organizes narrative around the development of a protagonist and the impulse that elaborates an inclusive pattern of simultaneous relationships'.[18] This is why when the narrative initially shifts to Nell on her journey Dickens may at first seem to present nothing more than a kind of intense, deeply carved 'Meantime . . .'.

From the first, however, there are warning signs. Nell herself is keenly aware of her loss of relations with the others, and the single-focused, sequential story of her picaresque journey increasingly comes to signify through the overshadowing absence of its multiplottedness. When she departs, this lack initially refers only to her withdrawal from the affairs she has left behind in London. Then she meets people on the road. And whether with the Punch men, Codlin and Short, who begin conniving at returning Nell and her grandfather to whomever might be missing them in London, or later with Mrs Jarley, when Nell stalks the Edwards sisters because she wants so badly to be part of those other lives but instead gets brutally sidetracked by her relapsing grandfather's conspiring with thieving gamblers, Nell repeatedly finds her story intermingling with others', the reader feels the creeping encroachment of the missing, larger multiplotted form, and then Nell—with the reader in tow—walks right out of it all over again. The significance is definitive. Where the city can sustain a multiplotted form because people's paths there are crossing and recrossing whether they know it or not (and Dickens's aim is for readers to know it as something they can never fully know), Nell and her grandfather's intercity walk cannot sustain the multiplotted form that would spring up around them. The long-distance foot-journey means the connections keep breaking.

This disconnect is—not forgetting that Pickwick is listening—a historical view of intercity walking that registers its transformation by a public transport revolution. In a complete reversal of the historical function of the road, Nell and her grandfather's pedestrian mode of intercity journeying un-networks them. And not only them alone, but all thus going it alone.

'Which way?' (3.156) Nell demands after she and her grandfather sneak out of the home they no longer own into the street and their wandering escape begins (with money enough for coach fare still in her pocket). This time, unlike when she first appeared to Master Humphrey and the reader, Nell is not actually asking for directions. They are not headed somewhere in particular. Like Pickwick and Sam, they are simply going. Nor is her question especially meaningful because the answering silence of her road companion makes it 'plain that she was thenceforth his guide and leader' (3.156). That's hardly a surprise. The sickly grandfather presents a near-Gothic version of sidekicks like Sancho Panza or Sam Weller. Rather, like Pickwick's question 'Where shall we go next?' Nell's 'Which way?' implicitly refers back to the means and meanings of her going. No 'Cab!' to the Golden Cross Inn and a speedy stage coach will undergird this trip by shanks's mare. Nell's 'Which way?' indexes a mode of long-distance journeying in which she will relentlessly and perpetually face solo decisions about this way or that, about 'direct[ing] their weary steps' (4.176) both in the broadest figurative sense of taking steps and also in the literalist, narrowest, physical sense of simply orienting the body and walking on. This lonely which-way relation of self to road is taking hold, as Phiz illustrates it, positioning Nell awkwardly askance her grandfather, foot forward (Fig. 12).

Wanderers though these two may be, from here on the course of the novel might be said to be set. The break from London initiates a journey from which unfolds a tale about, on the one side, their disconnection from a networked community and about, on the other, the netting of that communities' members.

Approaching the story's second cycling back to the action in London, the first long segment of Nell and her grandfather's foot-journey culminates at the horse races in something like a symbolic summarization of their lonely outsider nomadism. If on the long march to the races 'often a four-horse carriage, dashing by, obscured all objects in the gritty cloud it raised, and left them, stunned and blinded, far behind' (4.197), the racetrack itself epitomizes a horrifying, exclusionary gathering of people around horses speeding around in pointless circles to precision timing. When Nell, begging, memorably offers some flowers to a woman in a carriage there, her outcast status gets juxtaposed with that of a gypsy woman competitively hawking fortunes and with the woman in the carriage, who turns out to be a

Figure 12. They are off. Nell and her grandfather departing London. Courtesy of the Department of Special Collections, Young Research Library, UCLA.

prostitute. And while there is much all three marginalized women share, Nell's walking journey differentiates and defines her. Hers is the story of the trudging female foot-passenger—also offering up flowers—whom Pickwick is seen speeding past in Phiz's illustration of 'Bob Sawyer on the Roof of the Chaise' (Fig. 6). Polishing off this segment of her trek with an emblematic image, Nell and her grandfather cut across and away from the speeding circle that is churning everybody else: 'The bell was ringing and the course was cleared by the time they reached the ropes, but they dashed across it insensible to the shouts and screeching...' (4.200).

When Dickens then sweeps the reader back via a new chapter (the twentieth) to London and leaves behind Nell and the grandfather's plot, what's crucial to grasp is that the narrative nonetheless continues an exposition that illuminates both sides of this divided story. Just as the scene at the horse track that the reader just read partly took its meaning from snapping the plot threads of a multiplot form, from paths terminally diverging, back in London it's just the opposite.

Interrupting his worrying about Nell and her grandfather, Kit suddenly remembers that he must race off to keep his one-week appointment (timed to the minute) to hold the pony of the clubfooted Mr Abel Garland and his clubfooted son. After having completed this act of astonishingly responsible punctuality (regarded by all as a joke when the story had previously left him) and relapsed again into worrying about 'his late master and his lovely grandchild, who were the fountain-head of all his meditations' and 'casting about for some plausible means of accounting for their non-appearance': 'lo and behold there was the pony again' (5.205). The Garlands have raced on ahead to Kit's home to interview Kit's mother and hire Kit into their family's service managing this wayward pony (Whiskers). In a simple way, what's going on plotwise is that, having established villains to spare, Dickens is bulking up a troop of middle-class saviors. At a much more acute level, however, readers find themselves immersed inside a plot—cast in a crosslight by Nell's disconnection—that is about the meaningful recounting of the crosscutting and the intersecting, the whisking, between individual plot-paths in London that is the veritable lifeblood of the multiplotted novel.

The contrasting modes of mobility create contrasting modes of plotting, and the street scenes especially profile the differences. Consider, for instance, the seemingly arbitrary and bizarre episode in which Dick Swiveller bewails his orphaned state as he drunkenly reels 'after a sinuous and cork-screw fashion, with many checks and stumbles' down a crowded London avenue (5.217). In a surprising twist, Swiveller unexpectedly discovers along with the reader that he has completely blotted out that he is walking along with Quilp, 'who indeed had been in his company all the time but whom he had some vague idea of having left a mile or two behind' (5.217). The rediscovered Quilp offers on the spot to be a second father to Swiveller. Their urban street encounter, as Phiz illustrates it, visually quotes the conjoining of urban pedestrian bodies that sculpted the tale's opening's illustrated letter 'N' ('Night is my time for walking...'). In this scene, as in the inhabited letter, the connectedness of bodies is neither self-evident nor epistemologically neutral. In the paved warren that is London, where everyone passes and repasses each other, adjoining plots may adjoin, but the problem lies in knowing such connectedness.

Compare that consciousness-raising scene of urban impartibility with Quilp's unexpectedly flitting right past the orphan Nell in some quiet, small town's empty street in the next segment of her journey. Quilp will never come anywhere close to seeing Nell again. His abrupt, brief appearance might at a glance read like some kind of clunky reminder of the ongoing action in London, perhaps even as a minor narrative mis-step to be chalked up to the uncorrectable rush-job engendered by serial production. But Quilp's flash eruption into Nell's single-focused, sequential plot thread in fact actually appears completely consistent with it as such. Of course Quilp materializes as if he had 'risen out of the earth' (5.246): when the narrative follows Nell, readers have little idea, except as filtering through her limited view (and Dickens never disregards her youth), of the surrounding, compounding, multiplotted world where many of the singularly crucial events of her life occur in her absence. How did Quilp get there? Why? Readers will never know. This is part of the scene's point. As Quilp rushes past Nell tucking herself into the corner of a Gothic gate (Pugin's portal to another time foreshadowing Nell's respite to come), Dickens at once both records Nell's quailing through her startled eyes and in so doing chronicles it against the absent vantage of the city she has left, where following the development of the various separate plot trajectories one might have comprehended this non-intersecting criss-crossing of paths.

But then Nell would be crucially networked in. She is not. Instead Pickwick is listening; the railways are humming. Here is Quilp; he is turning back toward her beckoning—'To her? oh no, thank God, not to her...another figure—that of a boy—who carried on his back a trunk': 'Faster, sirrah! faster!...you creep, you dog, you crawl, you measure distance like a worm. There are the chimes now, half-past twelve' (5.246–7). That fast intercity networking machine, the London night coach, is due precisely at one o'clock. 'Come on then...or I shall be too late. Faster—do you hear me? Faster' (5.247). As Quilp rushes right past his sought-after Nell, readers may feel their first trickles of the irony to come inflecting Quilp's cries of 'faster' and his worries about being 'too late'. These are hints that perhaps no speed will be quick enough for the London characters to catch up with Nell, that maybe even the slower she crawls along her lonesome path, the less this is likely.

'Be in time, be in time, be in time!' Mrs Jarley calls out repeatedly at the conclusion of this segment of Nell's journey, closing Chapter 32 (6.276). Mrs Jarley is advertising to potential patrons that she is packing up the waxworks and moving on. As the ringing, closing imperative of Nell's initial sojourn with Mrs Jarley, however, Mrs Jarley's mantra-like command to 'Be in time' gathers an almost philosophical density as it reverberates inside Master Humphrey's clock and booms out across the plot's immediate switch back to London. The question re-echoing in that many-sided London surround is, Will anyone be in time? Is there anyone who will be in time? And, what will it involve, this being in time?

Enter the character known initially only as the single gentleman, who will twice command failed coach pursuits after Nell and her grandfather. Or to be more accurate, since the single gentleman's entrance occurs in the multiplotted domain of London, and Dickens really is telling his readers something about the meaning of being in time, enter first the omniscient narrator showing off that narrator's dazzling transportive power to weave retrospective narratives into an ongoing inclusive constellation of simultaneous activity into which this single gentleman enters: 'As the course of this tale requires that we should become acquainted, somewhere hereabouts, with a few par-ticulars connected with the domestic economy of Mr. Sampson Brass... the historian takes the friendly reader by the hand, and spring-ing with him into the air, and cleaving the same at a [great] rate... alights with him upon the pavement of Bevis Marks.... The intrepid aero-nauts alight before a small dark house, once the residence of Mr. Samp-son Brass... [where there is a] bill, "First floor to let to a single gentleman"... tied to the knocker' (6.277). Such rapid commuting across time and space is tellingly disorienting. Is this house already no longer Brass's? 'Once the residence'? Readers cannot be immediately sure because the novel spins both its own past present moment and those of its characters' from retrospective narratives. Brass's current home it is though, and, in the weekly number, Dickens conducts the reader through Dick Swiveller's hiring as the Brasses' clerk, the addi-tion of two new characters (Sally Brass and the small servant), and then on, finally, to the single gentleman's arrival, whose dovetailing into the multiplot action is no more accidental than the narrator's earlier men-tion of the ad for a 'single gentleman'.

Predictably, this single gentleman's introduction of one-way attempts to cross the novel's central plot divide from London to Nell will hardly unfold in any kind of balanced way on either side of that divide. Quilp may simply erupt into Nell's single-focused narrative, but because, taken as a whole, readers are reading an omnisciently told, multi-stranded novel—albeit one unraveling around the thread pulled from it—the attempt to bridge the disjunction between Nell on her foot-journey and the ensemble of London characters as seen from within their multiplotted narrative by definition will not unfold from the point of view of a single character's plot thread. It will unfold across the fabric of their interconnected time and space. The single gentle-man may, at first, see himself as on a single-minded solo chase, but, as he discovers, he is mistaken. His chase implants him in a community, and it will unfold across that community's synchronous, intersecting activity.

As the aeronautical historian announces, the way to knowing that community comes partly through weaving together its members' ret-rospective narratives. This is because of the momentous fact that the knowledge of separate simultaneous happenings does not actually get produced in the moment of its simultaneity. It must be pieced together through the sequencing of retrospective narratives. Just as simultaneity at a distance is a laborious convention of synchronizing clocks, the narrative correlating of separate events represents a mode of knowing that is a convention, specifically a convention dependent on commu-nicating first this and then that as potentially time-coincident.[19] It is a serial process.

And *The Old Curiosity Shop*'s serial publication especially made the sequential nature of the knowledge of simultaneous activity palpable. The serial readers really did experience the separate synchronous plot threads alternating in time. For them, serialization made it collectively and objectively true, and not merely an aspect of an individual's read-ing, that the reader was unable to hear—must wait to hear—about the concurrent activity of other characters.

By contrast, once upon a time there was a novel—one might pre-tend to reminisce—a novel where a plurality of ongoing activity appeared wonderfully to be happening everywhere simultaneously all around. Adventure after adventure in this novel came together around a hero, the stage-coaching network for which he stood, and the novel (also serialized), all of which bore the same name. Sustaining this nov-el's perpetual sense of encountering a shared world everywhere in plural motion required, however, that its omniscient narrator did not actually trace any very extended threads of the characters' separate storylines. That would have meant alternating between sidelong-

running retrospective narratives, and what quicker way for this
novel—it is *Pickwick*—to spoil its reader's exciting sense of immediately encountering a shared world in plural motion than to show that
the knowledge of that pooled, communal action might need to be
sequentially assembled in relation to a future selective of a combina-
tory, backward view of the many-threaded past. But separate simultaneous action cannot be made simultaneous until after it happens; one
has to do reconstructive work to make it so.

This simple fact has far-reaching implications upon which *Master
Humphrey's Clock* shines some light. With the entrance of the single
gentleman in *Shop*, the narratological re-assembly process that
would bring its two main disconnected retrospective narratives
together becomes the story's conspicuous focus. Within the story-
world, the single gentleman begins reconstructing Nell and the
grandfather's movements to place them in his present. He has traced
Nell and her grandfather to the horse races, and he begins methodically interviewing every traveling Punch man that arrives in
London. The narrative dilates upon his locating Codlin and Short.
Plot lines converge, and though these two characters no longer
know where Nell and her grandfather are, they do know—the
multiplot seems generatively inclusive—a character previously
introduced who might. Jerry 'wot keeps a company of dancing-
dogs, told me in a accidental sort of way, that he had seen the old
gentleman in connexion with a travelling wax-work, unbeknown
to him' (6.306).

So do Nell and her grandfather stunningly reappear within the
multiplotted half of the novel inside a flashback (Fig. 13). Returning
the reader to a moment never before known to have happened, this
flashback retrospectively creates a chain of connections that holds out
the promise of the convergence of the London crew with Nell and her
grandfather on their journey.

What follows next in the story? What does Dickens relate after the
single gentleman has in hand the clue that will launch his pursuit of
Nell? Certainly not the single gentleman's pursuit. In the next chapter
the narrative immediately cuts away with the omniscient narrator. The
aeronautical historian now very calculatedly commences interweaving
the London characters' paths. The multiplot's subdividing-and-
aggregating form again explicitly asserts itself, and the narrator goes
back in time to pick up the tracks of another of our co-starring

Figure 13. A Backward Glance. Nell and her grandfather observed. Courtesy of the Department of Special Collections, Young Research Library, UCLA.

characters: 'Kit—for it happens at this juncture, not only that we have breathing time to follow his fortunes, but that the necessities of these adventures so adapt themselves...to call upon us imperatively to pursue the track we most desire to take—Kit while the matters treated of in the last fifteen chapters were yet in progress, was, as the reader may suppose, gradually familiarising himself more and more with Mr. and Mrs. Garland...' (7.1). Far from dismayingly hearing Dickens clumsily cranking up the novelistic machinery, readers should feel here that they really are 'breathing time'. As readerly assurance has it—'the reader may suppose' because the reader has presuppositions—the omniscient narrator will meticulously go back to bring Kit's plot forward to its connection to the single gentleman's.

This perfect coming-together of Kit's plot with the single gentleman's is, I am suggesting, a purposeful set-up.

'Christopher...I have found your old master and young mistress.'

'No, sir! Have you though?...Where are they, sir?...'

'A long way from here.... If we travel post all night, we shall reach there in good time, to-morrow morning.' (7.17–18)

As Kit informs the single gentleman, he cannot go. The grandfather hates the sight of him for reasons unknown. This serves to remind readers that Quilp misled the grandfather into thinking Kit betrayed him, and the seeming hitch sets in motion more multiplotted braiding. Kit will race through 'the crowded city streets, dividing the stream of people, dashing across busy roadways, diving into lanes and alleys' (7.19) to drag his mother from church and the margins of the story to go in his stead. There at the church too is Quilp, who the reader was just nudged to remember. He is spying on Kit's mother. As will be revealed later, he has just discovered from Dick Swiveller that the single gentleman has called upon Kit, thereby linking the two of them for the first time for Quilp. In other words, other tributary plots were interweaving elsewhere; the retrospective plot strands are all beautifully intersecting. . . .

The first coach race to reach Nell is thus not framed merely in terms of rescuing her from her harrowing journey. Indeed, as far as readers know, at this point Nell is still teamed up happily enough with Mrs Jarley. Rather, as the narrator describes the post chaise departing from the point of view of Kit with 'tears in his eyes—not brought there by the departure he witnessed, but by the return to which he looked forward' (7.24), the race is on to try to recover Nell into the extended community in London whose identity as a community turns out at its lowest common denominator to depend on the capacity to produce such convergences out of the many characters' mutually constitutive retrospective narratives.

All this becomes much clearer—Dickens is essentially smacking his readers over their novel-reading heads with it—if one turns to the pivotal fact that the single gentleman arrives 'too late' (10.199). Therein lies the tragic meaning of Nell's departure. The devastating compositional logic at this novel's core is that Dickens is creating a retrospective narrative composed in its central alternating plot structure of two different retrospective narratives that fail to converge.

In the wake of this first coach pursuit, the centrally divided plot becomes more divided than ever. Where previously Nell's journey was set against the web of activity she left behind in London, now it is London plus the pursuing stage coaches stretching futilely out from it for her.

It is not that the pursuing coaches extend the ambit of the city. Nell and her grandfather, as well as the London crew, have left London.

Rather, the long-distance coaches extend an intercity network, and (as we saw in Chapter 1) this network forms, separate from an urban collective's intense and dense physically proximate arena of movement, a basis for organizing a networked time and space. So, even though a simple physics of competitive mobility would seem to ensure that this speeding coach should overtake these two plodding pedestrians, in fact not despite the swift coach but because of it, the single gentleman will not find Nell and her grandfather in time or space. One cannot find someone in a network who is out of it.

Just as historically Nell and her grandfather's long-distance walk presents an exiling form of solitary travel, so conversely—and remember that Pickwick is still listening to this tale—the failure of this stagecoach pursuit presents passenger transport as binding the characters together beyond London. Having raced sixty miles to some town with Kit's mother, the single gentleman finds in addition to Mrs Jarley: Quilp. The narrator explains: Quilp, having 'traced them to the notary's house [more plots crossing]; learnt the destination of the [single gentleman's] carriage from one of the postilions; and knowing that a fast night-coach started for the same place, at the very hour which was on the point of striking, from a street hard by, darted round to the coachoffice without more ado, and took his seat upon the roof. After passing and repassing the [single gentleman's] carriage on the road, and being passed and repassed by it sundry times over the course of the night...they reached the town almost together' (8.65). Arriving together with these stage coaches that arrive together is also, at another level, the London multiplot and, coach racing coach, its characters' retrospective correlativity in a network generative of their simultaneous time across long distances. The single gentleman, encountering Quilp instead of Nell and her grandfather, accuses Quilp of 'dogging [his] footsteps' (8.63). Quilp—who long precedes him in the hunt for Nell—mostly (but not completely) disingenuously counters: 'how do I know but you are dogging *my* footsteps' (8.64). In fact, neither is dogging the other's 'footsteps'. Nell is the one 'printing her tiny footsteps' across these pages (5.225). But in a larger sense these quarrelers also have it right. Invoking autonomous walking accurately expresses their frustrated sense of mutual indivisibility.

Contra Chesterton, Dickens's artistic idea was not quite only that 'the good powers...should set out in pursuit of one insignificant child, to repair an injustice to her...and should all—arrive too late'. As the first stage-coach pursuit makes clear, a hero, a villain, and a fringe character all arrive too late to find Nell. And whether they are planning to assist her, exploit her, or just along to lend a hand, what's actually not at stake for readers is Nell's welfare. Any reader who believes that it would be worse for Quilp to find Nell than for her to keep slogging on alone with her grandfather hasn't been paying attention to Mrs Quilp or woken up—as Nell terrifyingly has—to the grandfather.

At stake in the coach pursuit is the attempt to re-include Nell and her grandfather in a community from which their foot-journey unnetworks them and that community's confrontation with its limits. In this respect, it's worth pausing to admire that final dash of Chesterton's in his description that they 'all—arrive too late'. That dash almost reads like a tiny path one eye-travels only to be mimetically pulled up short, something like the leash that chokes the dog just shy of Quilp as he wallows in self-delight over his deep-laid plots. And the dash reads as its verb. *Dash*: ' "This is the place!" cried her companion [the single gentleman], letting down all the glasses. "Drive to the waxwork!" The boy on the wheeler touched his hat, and setting spurs to his horse, to the end that they might go in brilliantly, all four broke into a smart canter, and dashed through the streets with a noise that brought the good folks wondering to their doors and windows, and drowned the sober voices of the town-clocks as they chimed out half-past eight' (8.58). *Dashed*: 'This discouraging information [that the town was eight miles off] a little dashed the child, who could scarcely repress a tear as she glanced along the darkening road' (5.238). Dashes dashed.

Digressive meditations on dashes aside, however, the failure encapsulated by the arrival 'too late', toward which the dash futilely stretches, presents more than a missed meeting in this novel. It cuts to a fundamental ligature of imagining others' separate activity in 'Meantime' relation. After the usual chorus begins crying 'Where is she?' the narrator steps forward to identify exactly this ruptured simultaneity as at issue: 'What would he [the single gentleman] have given to know, and what sorrow would have been saved if he had only known, that at that moment both child and grandfather were seated in the old church porch, patiently awaiting the schoolmaster's return!' (8.60). Because readers left Nell and her grandfather sitting on that church porch, the narrator's speculation is disclosing no new information. Rather, it slaps readers with the simultaneity failing to be achieved. The broken spatiotemporal bridge reverberates through the accumulating conditionals ('What would...what would...if...only'); through the assertion that the simultaneity is a missing state of knowledge ('if he had only known...'); and through the poignant detail that Nell and her grandfather themselves only anticipate the result of someone else's strictly local simultaneous activity ('patiently awaiting the schoolmaster's return'). Overshadowing all, the alternating plot structure relentlessly persists. Nell and her grandfather may remain alive, not even so very far off 'at that moment', but Dickens dramatizes that that does not quite suffice to connect the others with them.

For her London-based pursuers, the disconnect straightforwardly means that upon their arrival they make 'all possible inquiries that might lead to the discovery of the old man and his grandchild. But all was in vain. Not the slightest trace or clue could be obtained. They had left town by night; no one had seen them go; no one had met them on the road; the driver of no coach, cart, waggon, had seen any travellers answering their description; nobody had fallen in with them, or heard of them' (8.66). For readers, this impasse causes Nell and her grandfather's journey onward to become, after the fact, newly meaningful for its having foiled the stage-coach pursuit. In other words, the coaches' belated arrival brings home belatedly, retrospectively, the severing divide forged by the previous segment of Nell's journey, which includes her flight from Mrs Jarley's; her canal trip into the unnamed city readily identifiable as Birmingham; her stopover with the furnace man; and then her onward north-westerly tramp through the Black Country's industrial wasteland, until she collapses on the road at the schoolmaster's feet and is forwarded by wagon to the church porch.

Conventionally, the narrative logic of sequencing retrospective narratives—here the switching between Nell and the London ensemble—would dictate a selective process of narration mounting to an endpoint that allows the reader retroactively to correlate, however incompletely and discretely, the characters' synchronic activity together in time and space. Exactly the opposite principle is in effect here. Dickens sandwiches the relaying of Nell and her grandfather's week-long journey between the departure and the arrival of the pursuing coaches, jumping back a week to explain how it will be that, when the speeding coaches arrive, the single gentleman will have been on a 'wild-goose chase' (9.111). On both sides of the central plot-divide, this story looks back and follows plots forward in time and space toward a retrospect explicative of their disconnected non-convergence.

Retrospectively, Nell and her grandfather's journey tests fully positive for that litmus of tragedy: irony. The segment begins: 'It behoves us to leave Kit for a while, thoughtful and expectant [watching the coach racing off after Nell], and to follow the fortunes of little Nell; resuming the thread of the narrative at the point where it was left some chapters back' (7.25). By chapter's end, Nell is in a frenzy: 'Up! We must fly'; 'There is no time to lose; I will not lose one minute. . . . Up! and away with me!' (7.31). Up and away? No time to lose? Where readers might at first perhaps worry about this plot turn, pondering its relation to the coach known to be coming, looking back they know the ironic truth. Nell and her grandfather are fleeing their own imminent rescue. What would Nell have given to know, and what sorrow could have been saved if she only could have known. . . . After fleeing Mrs Jarley's, Nell consoles herself: 'I have saved him . . . I will remember that' (7.32). As it falls out with poetic injustice, she has saved her grandfather from reunion with his once-beloved, long-lost brother and untold riches, while her own heroic, selfless marching onward only merits her all the more the rescue it precludes. Because the pursuing stage coaches, unlike the London she left behind, travel through the same region through which Nell and her grandfather also march, a spatial overlay creates a powerful echo-chamber for Nell's words and acts. The resulting irony registers that Nell and her grandfather's journey no longer merely disconnects them from the others. It indicates that their being disconnected is the connection between them.

As an aspect of this amputated relation, the full significance of Nell and her grandfather's canal ride now also heaves into view. Tremendous plot weight retroactively falls upon Nell's waterside decision to accept an invitation to embark on a barge with some boatmen: 'The child hesitated for a moment, and thinking . . . that if [she and her grandfather] went with these men, all traces of them must surely be lost at that spot; determined to accept their offer' (7.33). She is only too correct. On the canal, 'nothing encroached on their monotonous and secluded track' (7.33).

Nell and her grandfather float right off the road map. The canal ride temporarily absorbs these foot-travelers into an inland transport system with its own independent physical infrastructure. Locks, inclined planes, aqueducts: revolutionizing an ancient process of making rivers navigable—especially through straightening by cuts—the dramatic rise of the canals in the eighteenth century succeeded in proving for the first time that a whole new kind of inland long-distance transport system could be engineered. In this, the canals prefigured the railways. (The canals' wildly lucrative success at shipping bulk goods—not dreams of paying passengers—motivated the initial building of the steam rail network.) From the glorious opening of James Brindley and the Duke of Bridgewater's famous canal in 1761 up to its precipitous decline in the face of the railways in the nineteenth century, over 3,000 miles of waterways reticulated Britain.[20]

By the 1820s the main canal arteries look, to my eyes at least, something like a giant figure 8, with Birmingham roughly center spot (Fig. 14), the network held down at each corner by one of England's major rivers: the Thames, the Severn, the Mersey, and the Trent. Nell and her grandfather's route, stretching from Quilp's dilapidated wharf on the busy Thames to the fully industrialized wharf in Birmingham, wends right to the heart of this canal network.

Thus goes the journey of Nell *Trent*: where Nell might have seemed to have taken her surname from one of England's longest rivers, in fact she merges with the destiny written into her last name when she heads down a polluted freight waterway remotely channeling that river's water. So much for her and her grandfather's dreamy idylls, which the narrator undercut as they were forming, about 'travel[ling] afoot through fields and woods, and by the sides of rivers', 'free and happy as the birds' (3.153–4). Uplifting sentimental pastoral adages, like the one sung by Swiveller, about how 'life like a river is flowing' (9.121), are just that: uplifting sentimental pastoral adages. Fate, at whom Swiveller shakes his fist, cuts the legs out from under everyone; it is an 'accumulation of staggerers' (6.285).

The canals, like the Warwick–Birmingham on which Nell and her grandfather journey, were not primarily for passenger transport. By the 1820s and 1830s, the canals' failure to revolutionize passenger transport must even have seemed particularly glaring. Short-distance urban commuting on river arteries thrived from the Clyde to the Thames. The boatmen had had a decades-long lead over land travel in using steam power for local commercial passenger transport, beginning with the *Comet* on the Clyde in 1812, the year of Dickens's birth. Coastal packets plied successfully between town ports. And yet—despite 'fly-boats' (still too slow) and an extensive waterway network,

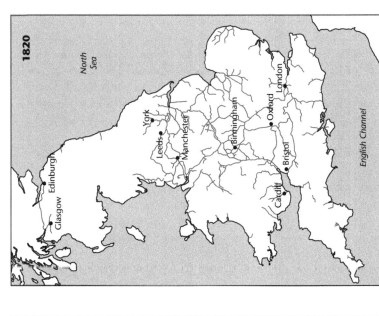

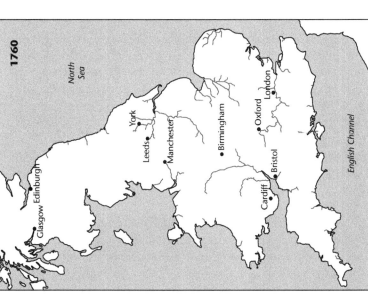

Figure 14. Inland waterway maps, 1760 and 1820. Reproduced from Charles Hadfield, *The Canal Age* (Newton Abbot: David & Charles, 1968) © Charles Hadfield.

the canals had failed to transform or systematize intercity passenger transport.

They carried industrial freight. Not surprisingly, Nell's ordeal at the boatmen's hands, in which they force her to sing all night, has thus been read in economic terms and as of a piece with the industrial hell in Birmingham and the Black Country into which the barge delivers her. In this view, Nell's compulsory singing performance joins a whole series of renderings of public entertainment in the novel—Punch, Jarley's waxworks, Astley's. All reflexively mark, in different ways, the commodification of Nell, this novel (*Shop*), and the author making it pay.[21] On this axis, Nell's canal trip also lines up with her first ride in a jolting farmer's cart and her last in a crawling stage wagon, 'the child comfortably bestowed [*sic*] among the softer packages' (8.53). Both those lifts treat Nell as freight, and both tellingly drop her off at cemeteries. In relieving Nell's aching body, they consign it to its state of dead objectness. The same may be said of Mrs Jarley's waxwork's Common Stage Wagon: Mrs Jarley and her driver somewhat too literally weighing up Nell and adding her to the stock of wax figures.

But these twin threats ultimately sustain the primary distinction that the story is depicting Nell's living, dying body journeying. Thus: '"We shall be very slow to-day, dear," she said, as they toiled painfully through the streets; "my feet are sore, and I have pains in all my limbs…"' (7.44). Nor is Nell an industrial slave. Miss Monflathers's belief that Nell should be at 'work, work, work', 'assisting…the manufactures of your country' and 'improving [her] mind by the constant contemplation of the steam-engine' (6.269), envisions Nell in a different story than the one that Master Humphrey offers in which Nell walks, walks, walks. Nell and her grandfather never get enrolled in the industrialized netherworld. They pass through it as travelers.

Re-examining, then, the canal episode in terms of their journeying, what's striking is how the boatmen betray the relationship of conductor to passenger. As Nell and her grandfather float on through the night, the grandfather falls asleep and Nell falls into a reverie induced by the journey. 'How every circumstance of her short, eventful life, came thronging into her mind as they travelled on!…all the fancies and contradictions common in watching and excitement and restless change of place, beset the child' (7.35). But then: 'She happened, while she was thus engaged, to encounter the face of the man on deck' (7.35). And, as a result of this totally insignificant interaction, that is, without

any provocation, the boatman, who is viciously drunk, commands
Nell: 'Let me hear a song this minute . . . Give me a song this minute'
(7.35). No need for this driver to attend to timing the minutes of this
slack trip. Instead he invades the freedom that the public transport
system ordinarily grants passengers to remain otherwise preoccupied.
The boatmen vilely force 'the tired and exhausted child' (7.36), who is
'but poorly clad' (7.35), to spend the night 'singing the same songs
again and again' (7.35). They accompany her with an incoherent roaring that 'rent the very air', creating a 'discordant chorus as it floated
away upon the wind', and, as the narrator reports, 'many a cot-
tager . . . hid his head beneath the bed-clothes and trembled at the
sounds' (7.36).

If this trip sounds haunting, that may be because it is haunted. Harry
Hanson, the canal boatmen's historian, apparently unaware of Dickens's confirming portrait, not only pegs the intensification of public
disapproval of canal boatmen to the moment of this novel's publication, 1839–41. As Hanson also recounts, the lightning rod for public
repulsion was a gruesome and sensationalized rape and murder by a
three-man crew (barbarically drunk) of a female passenger, who was
overheard crying out for help for much of the ghastly night's journey
in June 1839.[22] There can be little doubt that Dickens, along with most
of his first British readers, knew of this notorious crime (which took
place on the Trent and Mersey) or that it haunts Nell Trent's canal
journey (on the nearby Birmingham and Warwick). It immediately
became a staple feature of popular crime anthologies. One account
appears in *Chronicles of Crime*, illustrated by Hablot Browne as 'Phiz',
published the same year as *The Old Curiosity Shop* (1841). As Hanson
reports: 'What was particularly horrifying about this [crime] was that a
woman had been attacked and killed in a public conveyance by the
very people who should have been ensuring her safety and welfare. It
was as if the driver and guard of a stage coach had raped and murdered
their only passenger.'[23] These are exactly the terms of comparative passenger transport to apply, I believe, to Nell's canal journey.

Pickwick versus Trent: transportopia, transportality. Opposing *Pickwick*'s comedy is not merely a mirroring tragedy. It is the recognition
that the paradise previously hailed was already lost. And more than
anywhere else inside *Shop* at the close of the canal episode the story
brushes up allusively against *Pickwick*. As the Dickensians who helpfully mapped out Nell and her grandfather's journey first observed,

when Nell and her grandfather dock in the unnamed town, which readers readily recognize as Birmingham, their journey directly overlaps with Pickwick's trip there to see the wharfinger Winkle.[24] Some doubts exist about whether they land at the same wharf, but what matters here is that the passage describing their arrival echoes Pickwick's:

> They had for some time been gradually approaching the place for which they were bound. The water had become thicker and dirtier; other barges coming from it passed them frequently; the paths of the coal-ash and huts of staring brick, marked the vicinity of some great manufacturing town; while scattered streets and houses, and smoke from distant furnaces, indicated that they were already in the outskirts. Now, the clustered roofs, and piles of buildings trembling with the working of engines, and dimly resounding with their shrieks and throbbings; the tall chimneys vomiting forth a black vapour... the clank of hammers beating upon iron, the roar of busy streets and noisy crowds... announced the termination of their journey. (7.36)

As the clanking of iron and the working of engines are once again foretelling, the railways are coming. But if the power of steam and coal epitomized by industrial Birmingham will accelerate the immortal Pickwick in his coach into a world-altering passenger transport network, that same system's time-warping power alienates the downtrodden from their own present: 'The child and her grandfather, after waiting in vain to thank them [the boatmen], or ask them whither they should go, passed through a dirty lane into a crowded street, and stood amid its din and tumult, and in the pouring rain, as strange, bewildered, and confused, as if they had lived a thousand years before, and were raised from the dead' (7.36).

Walking into town might at least have preserved their independence. Arriving by a defunct cargo network, a dead network unnetworks them utterly. For Dickens's 1840s middle-class readers, the rail link between London and Birmingham, opened in September 1838, had made the 110-mile journey, which essentially kills Nell, an easy five hours. Dickens, who first travelled that rail line in November of 1838, made the rail trip again right before the first day of *Clock*'s publication, following his usual practice of clearing out of town. 'All right/In great speed/Off!' he scribbled to Chapman and Hall inside the proof of the first part's wrapper. Re-read sequentially, instead of wholesale descriptively, as a tiny narrative beginning 'All right', followed by the complication of 'In great speed', that unhappily ends 'Off!', it was a more apt epigraph than he could yet have known.[25]

Having sent Nell up the canal into Birmingham, Dickens has done all he need do to scissor Nell and her grandfather from the first coach pursuing them. Already he is germinating a second, final failed coach pursuit. The reader is in for another chase arriving—conclusively this time—'too late again!' (10.199), with Nell irrevocably 'Dead' (memorandum) and the grandfather, dying, completely addled and unable to recognize them. Where the first failure mapped the network's boundary, exposing the boundedness of those within the system and the disconnect of those without, this second failure kills any notion that extending inclusivity might offer some kind of a solution. On the contrary, Dickens will now expose how the networked passenger transport system, which in *Pickwick* was shown to hold community together by offering access to individuals interchangeably, also defeats community for individuals on those same grounds.

As not only its readers know, but also a great many who have never read the story, this story's end focuses on Nell's death. There are early intimations of her mortality, such as the death of the little scholar, but her doom gets definitively inscribed somewhere in her ferrying along the latter-day Styx into the Hades that is Birmingham, where she receives from the furnace man the two coins she needs to pay Charon in order to be released from her mortal coil. Or is it, more drearily, simply that a deadly foot infection sets in on the road out of Birmingham through the wasted Black Country?—'she lay down, with nothing between her and the sky; and with no fear for herself, for she was past it now' (7.46); 'Poor child! The cause was in her tottering feet' (7.46); 'See here—these shoes—how worn they are. . . . You see where the little feet were bare upon the ground. They told me, afterwards, that the stones had cut and bruised them. *She* never told me that. . . . I have remembered since, she walked behind me, sir, that I might not see how lame she was' (10.206). Ultimately the onset of Nell's death sentence is supposed to be diffuse. It is the journey on 'paths never made for feet like yours' (7.43) that finally does her in, not some particular mishap during it.

When the single gentleman first arrives too late, the story raised this potential outcome: 'Oh sir! . . . why weren't you here a week ago?' 'She is not—not dead?' 'No, not so bad as that' (8.59). No, not yet, a reader may well reflect since the story has pretty clearly indicated that Nell at the church has reached her burial ground. Following Quilp's return by stage coach to London, Dickens previews the debacle to come in comic counterpoint. Quilp's journey has caused his immediate community in London to jump the gun and conclude from his sudden disappearance that he is dead, fallen from his wharf and drowned: 'gone, ma'am, to where his [crooked] legs will never come in question' (8.71). But he is not. He is still networked in. And after 'disappointing them all by walking in alive' (8.69), Quilp parodically absorbs his imagined demise. He holes up in ostensible isolation in his makeshift bachelor hall on his wharf, 'accountable to nobody for [his] goings or comings' (8.79)—'I'm dead, an't I?' (8.78)—while remaining totally involved in everyone else's comings and goings, 'a Will o' the Wisp, now here, now there, dancing about you always' (8.78).

Predictably, something like the opposite scenario unfolds in the novel's other half, which is the final segment of Nell's story (through to the sexton pointing her down into a gravelike pit in the church basement). Nell is quietly and grimly wasting away. As Philip Rogers has astutely laid out, in the four chapters describing Nell at the church, Dickens surrounds Nell with characters espousing a variety of different, even contradictory, viewpoints on death. There is that of the schoolmaster (who finds consolation in moving past it to its good influences on the living), the sexton (who perpetually engages in comical, competitive self-distancing and repression of it), and the bachelor (who welcomes it as 'another world, where sin and sorrow never came', 8.98).[26] To these main spokespersons for the ultimate silencer one could easily tack on three more: the little boy bewilderedly worrying about Nell's joining the angels while his schoolmates remain oblivious, the anonymous adult villagers who observe her dying with kind but measured and stable 'compassionate regard' (8.105), and Nell herself, who crafts a consoling gate to heaven out of the Gothic church, 'a place to...learn to die in' (8.86) (Dickens channeling Cattermole channeling Pugin).

All this talk of death is partly why—understandably, but nonetheless not quite correctly—so many readers have taken Nell's death to be the crowning subject of this novel. Both Nell and the story come to a standstill in a cemetery filled with homiletic voices, something akin to the texts strewn around on the tombs themselves. The novel feels positively epitaphic. Surely this must be it, The End. Send in the London crew to view the corpse and close the book. And yet, death in these chapters appears as a subject of speculation. It is abstract, and this is the point: it is abstracted from the novel's plot, divorced from it. The reader's

actual relation to Nell's death is different. For the reader, aligned with
the perspective of the multiplotted novel as a whole, Dickens brings
Nell, fatally ground down by her journey, to her last resting place and
then prepares for her death within her half of the story by surrounding
it with a slew of commentary all statically conjuring the same question: what does her dying mean, what does death mean?

Once one commands a view of this structure, it is but a small step
to discern that this same question—what does her death mean?—is
being visited upon the London stage-coaching crew (and, at another
level, upon the multiplotted half of the story). For them, as for the
reader, Nell's death is not a telos that encrypts virtually all that comes
before.[27] Her journey frames her death.

Readers in fact will not see Nell die. They approach Nell's death
along with, and through, the stage-coaching London community. In
negative terms, this approach meant deciding not—ever—to carry on
with Nell again. From that quarter (i.e. the novel's other half), a deathly
silence will reign. This silence signifies. By not having the narrator
return to Nell to accompany her encountering her death, by instead
keeping the reader with the multiplotted ensemble of London characters, Dickens does all he need do to set up the second coach pursuit
arriving too late and to frame Nell's death through the dropping out
of her breakaway journey.

So when Dickens leaves Nell in the church's basement with a grave-
digger (who points her down into an uncovered, disused, pitch-black
well, which unmistakably signifies her imminent destiny underground)
and sweeps his readers back to the plots in London, and specifically to
Dick Swiveller donning some black funeral crape to mark the death of
his romantic hopes (Sophia Cheggs having passed into matrimony), he
continues to illuminate both sides of his divided story. Swiveller attires
himself in what amounts for the reader to a non sequitur. And in a way
Swiveller's mourning band for this other girl, the wrong girl, this girl
replacing Nell whom Nell had earlier bumped out of the story, this
girl who is not really doomed (as Nell is), but rather entering into her
adult married life (as Nell never will), says it all. Back in London, Nell
now becomes so remote a figure that her absence hardly seems to
matter.

The London world to which she once belonged is spinning on
without her. Most obviously a plot sprung by Quilp to crush Kit
explodes. A legal intrigue in which Kit is framed for theft takes center

stage, holding in abeyance the story of non-converging journeys through a sequence of eleven chapters, including Kit's entrapment, his arrest, his imprisonment, his trial, his unjust conviction, and his exculpation, which, completely irrelative to Nell, then brings down the Brasses and Quilp. Fleeing, Quilp really does now fall from his wharf and the plot. Evocatively toppled by the absence of any human sounds that might guide his path forward through the fog, he is run over by a boat on the Thames highway and left a corpse washed up outside its channel markers, his last remark satirically granted that 'it might, for anything [he] cared, never be day again' (10.186). In the meantime, a new plot thread bound up with Kit's legal troubles also springs to the fore: the romance of Dick Swiveller and the small servant. A critical commonplace of this novel's many detractors is that its failures get partly redeemed by Dickens's sudden, unexpected branching off into the amazing creative force of mutual self-liberation that sprouts up between Dick Swiveller and the small servant. Rechristened 'the Marchioness', the small servant rises up at the novel's end from the Brasses' cellar like Nell's non-identical replacement twin 'appear[ing] mysterious from under ground' (6.288). John Bowen, Steven Marcus, Gabriel Pearson: all treat Swiveller and the Marchioness as the means to conclude their discussions precisely because they all quite rightly sense that relationship leading, as it were, outward and onward beyond Nell's story.[28] In a deeper way, though, that is the very purpose this relationship serves for this novel. Dickens means for readers to see the proceedings of the London community continuing on without Nell and her grandfather.

It is not that suddenly no one in London cares any more about Nell. For instance, in one of the few (there are three) mentions of her preceding the discovery of her location and the final coach pursuit, Kit in prison bewails the thought of Nell ever hearing of him there. But where for readers the urgent question preceding the first coach pursuit was 'Will they find Nell? Will they be in time?' right up to the departure of the second coach pursuit, the reader's question is something more like a befuddled: 'Wait, what about Nell?' In a spirited essay, Jerome Meckier has suggested that by prolonging and prolonging his focus on the resolutely local activity in London for seven weeks of the serial (numbers 36–42), Dickens provoked his audience to scan the various other characters' progresses for possible clues to Nell's fate, hoping against hope, even as they knew her death was inevitable.[29] But any such telegnostic searching, then or today, only confirms all the more that Nell's non-appearance concerns readers in its no longer being a direct concern of the characters, or apparent object of their plots. This multiplotted novel would seem to be falling completely apart. That is its design.

The overhanging question—what does Nell's death mean for this networked community?—is being answered. And, crucially, Nell's death cannot simply be pinned to the fact that the activity in London spins on without her, as if to say, facilely and myopically, life is only for the living. Nell is alive up until the morning they set out after her. Rather, when the active search for Nell halts, the community's composite relation to her disappearance begins to dissolve the difference between her having gone 'underground', in a figurative sense of becoming unlocatably alive, and her actually mouldering away under the ground. Quilp unwittingly drives this nail right into Nell's as yet unclosed coffin when he impatiently reminds his lackey Brass: 'am I to tell you . . . that I had a plot, a scheme, a little quiet piece of enjoyment afoot, of which the very cream and essence was . . . this old man and grandchild (who have sunk underground I think) . . .?' (9.149). In those parentheses, bending their curves here into a typographical pit, Nell and her grandfather sink into an underground that both refers to their parenthesizing from the London plots and at the same time blurs right into the fact that Nell, along with her grandfather, are headed underground in a much more earthly sense.[30] This is to say that gone is gone for these two goners, but not that the difference between whether Nell is merely gone or actually dead and gone does not matter to Kit or the single gentleman or Quilp or even to the novel's readers. On the contrary, Dickens is delineating how Nell's journey reveals how the community counterpoised to and defined against that journey relates to an individual human's death.

Unstopped by her dropping out, the networked community out of which Nell walked is equally unstopped by Nell's death. Put it this way: Nell disappears again, but this time she disappears the way everyone must eventually exit this network. The form of community construed in relation to Nell's foot-journey is constituted around the networking together of the living individual characters coming and going on their individual paths and around the coordinating of those characters' ongoing simultaneous activity, and for that very reason it also keeps right on going. The passenger transport system only grants communal value to individual lives considered in their circulating, present-tense aggregate.

This story's tragedy is not really the personal suffering that Nell endures, nor is the pivotal issue whether her dying appears awfully sad or laughably maudlin. There has been a century of delightfully withering scorn for Nell's martyrdom that stretches from Oscar Wilde's single, beautifully deadly, epigrammatic assertion that it would take a heart of stone not to laugh at the death of little Nell, through Aldous Huxley's ringing, stinging expressions of disgust at all the morbid sentimentality, to Gabriel Pearson's scalpel-like scholarly *coup de grâce* that left only 'Dick the liverer' living.[31] But a historical consciousness about Nell's fateful walk, along with a recognition that broadens the meaning of her death from her to the community it strikes, may help explain why so many of the novel's original readers anointed it one of Dickens's best.

Recalling that the previous stage-coach pursuit ironized Nell on her journey, framing her foot-journey inside a context of networked mobility unavailable to her, one might be tempted to suggest that Nell's ghost turns the tables. Dead Nell retrospectively ironizes the second coach pursuit by placing the community in the position of racing to re-include that ultimately unavailable drop-out, a corpse. The truth, however, is otherwise. Nell's decease precludes *her* re-establishing any ongoing, meanwhile relation to the community that she long ago left behind, but the same does not quite go for the community.

'Miss Nell—where is she—where is she?' Kit chokes out one last time to the grandfather when the London crew finally arrives (10.205). In a virtuoso display of control Dickens extends for pages the problem of providing any simple answer to that question, since Nell is after all both here and not here, there not being there. Partly a euphemism of endless sleep sustains her deathbed's opacity: 'Sleep has left me.... It is all with her!' (10.207). Mostly though the reader is kept reeling by destabilized verb tenses: 'Why dost thou lie so idle there, when thy little friends come creeping to the door, crying "where is Nell—sweet Nell?"—and sob, and weep, because they do not see thee. She was always gentle with children' (10.206). The grandfather is uprooting the dovetailing conventions of narrative retrospection, right down to the grammar distilling past from present. The resulting awkwardness does not hold the reader in suspense. It is clear what to expect. Rather, while the narrator focalizes the grandfather's senile disbelief to justify the surrealism, this subjective unreality becomes the real entranceway to Nell's death for the other characters and for the reader because it accurately renders their uncanny unilateral relationship to her corpse in simultaneous time and space.[32] The novel makes real for readers how, when the London community reconverges with Nell dead, their capacity to include her as contemporaneously there stands against her incapacity to contribute to a network capable of sustaining their shared time and space of simultaneous activity. Nell's death, which, framed by her journey, already foregrounds the community's amalgamating relation to individuals, exposes the non-reciprocity of this relation.

The deranged grandfather also epitomizes this non-reciprocity that obtains between the lone individual and a community organized around networking individuals' comings and goings. 'In' is 'in', whether an individual acknowledges it or not. Here is the grandfather: 'You plot among you to wean my heart from her. You never will do that—never while I have life. I have no relative or friend but her—I never had—I never will have. She is all in all to me. It is too late to part us now' (10.209). As with Nell, the Londoners both are and are not 'too late' for the grandfather. Confusingly, meaningfully so, the grandfather salutes family and friends he would deny knowing. His final scenes, following up and complementing Nell's death, show the community encountering an individual unable to recognize it and refusing to do so. The grandfather 'never understood, or seemed to care to understand, about his brother' (10.215), while his brother and the others organize all the more around him, carefully orchestrating, for instance, his absence from Nell's burial. Formally, their perspectives plays off the unavailability of his: 'he never told them what he thought' (10.216). The plot grinds into an iterative stasis, capable of swallowing in a paragraph the months until his death. The grandfather comes to do nothing but spend every day at Nell's grave, where he conceives his future solely in terms of continuing his solitary foot-journey onward with her: 'How many pictures of new journeys...how many visions of what had been, and what he hoped was yet to be...' (10.216). Up until the grandfather's joining with Nell in death, the walking journey the grandfather insanely pictures continuing with Nell offers the only counterforce that this character can muster against his re-circulation in the community.

Cattermole's illustration visually summarizes this road novel's final handling of Nell and her grandfather as seen from the London crew's perspective and the reader's (Fig. 15). 'HERE' is Nell, as Cattermole's pointedly abridged epitaph declares. And while the grandfather's fixed gaze reminds us that Nell creepily really is there, for the reader and the London crew that 'here' marks the networking, against Nell's foot-journey, through which their arrival establishes a 'here' insistently connected to other places.

A tragic paradox upon which Dickens builds the road tale *The Old Curiosity Shop* is thus that the public transport system only grants communal value to individual lives considered in their circulating, present-tense aggregate and not to any particular individual life, and yet this novel precisely tells a story about the preciousness of an individual life's lonely journey. Similarly, from the perspective of the individual reader, this novel's deepest irony perhaps emanates from its being a

Figure 15. The deluded grandfather ready to keep on walking, with Nell dead but 'here'. Courtesy of the Department of Special Collections, Young Research Library, UCLA.

novel, from its being an act of communication. In opposition to a passenger transport network, it memorializes. Through the second coach
pursuit and its arrival 'too late', Dickens has exposed the underside of
the aggregating—of the system's impersonal offer to carry Potts and
Slurks, Wellers and widders, all alike. Nothing, and certainly not any
individual's death, would seem to interrupt the collective, ongoing
momentum. The clock always keeps ticking away in railway time. In
the historical context of the passenger transport revolution's merging
of individual journeys and the resultant imagined, networked community, individuals would seem almost to become dispensable as
individuals.

III. *Clock* Strikes

The denouement of *The Old Curiosity Shop* extends past Nell's and the
grandfather's deaths back into the *Clock* frame. This denouement
begins to come into focus on the single gentleman's coach ride with
the surprise revelation en route that he is the grandfather's brother.
This revelation is then followed by a further thunderclap in *Clock* that
the single gentleman is also Master Humphrey. As he explains, he has
fabricated the tale's first-person opening and then narrated it omnisciently. Narratologically, these twists make for a complex conclusion,
to say the least. Far, however, from an absurdity best ignored—as it is
generally deemed and dismissed—Dickens is continuing to relate a
straightforward road story, one which the author might well be
expected to tell.

   Its own omniscient narration has just become the story's subject.
Cutting through all the usual critical hang-ups, Audrey Jaffe forged
ahead in grasping this. Rightly insisting on reading omniscience as
'the subject of narrative as well as a description of its form', Jaffe
splendidly sets out to analyze how revealing that the character called
the single gentleman is also the omniscient narrator of *Shop* illuminates 'omniscience...located...between a narratorial configuration
that refuses character and the characters that it requires to define
itself'.[33] Jaffe's intelligent discussion winds up sidetracked, however,
by a too-narrow focus on omniscience as knowledge. Her discussion revolves around the keyword 'curiosity' and reflexive scenes of
narration; she reads the story as about knowing about knowing

about the story. Jaffe's own sharp observation that omniscient narration combines a fantasy of mobility with knowledge thus wholly subsides into the latter. In reality, this novel stunningly puts those two together.

As I will argue, the ending leading back into *Clock* shows (amending that of the tale) that individuals in their networked community indispensably project the omniscient-like perspective—call it the view of an aeronautical historian—of their networking together. In its bare essentials, my contention is that Master Humphrey breaks out from the tale's (diegetic) story-world to reveal that he both exists within it and supplies a semi-omniscient viewpoint upon it. Moreover, this final, surprising plot turn makes complete sense in terms of his telling a tale about tragically broken relations of mobility. This argument brings into focus that this novel has two different endings: not only the familiar one in *Shop*, but also one in *Clock*, in which Master Humphrey directly invokes an omniscient perspective that unifies people through the (standardized) clocking of their simultaneous, separate, disparate activity. As suggested by the powerful sweep of this finale, in which an individual imagines an omniscient perspective emanating from a clock chiming, the tragic road tale with its hybrid first-person and omniscient telling belongs not just to Master Humphrey, but, as he announces, to everyone.

Shop is a road novel, and it introduces the message delivered through the postal network that prompts the final coach pursuit only to mow right over it. Mr Garland has received a letter from his beloved but neglected brother in a distant town about 'a child and an old man' that 'told there such a tale of their wanderings...that few could read it without being moved to tears' (10.192). This sudden, unexpected avenue of information (Mr Garland's brother turns out to be 'the bachelor' in the church village) not only informs the single gentleman and company in London where to go to find the grandfather and Nell, but also ostensibly marks their catching up, knowledgewise, with the other half of the divided story. Had Dickens been writing an epistolary novel a century earlier, this letter might have formed its closing catastrophe. Now a speeding stage-coach system bodily connects those people. The railways are humming. Intercity walks disconnect them. So Mr Garland fires off another note to his brother to confirm that the child and old man are Nell and her grandfather, receives quick confirmation, and, having wrapped up the plot hatched by Quilp to frame Kit that has been preoccupying everyone, Mr Garland suddenly takes Kit aside to explain that he again needs to get ready 'for a journey' (10.191): 'The place of their retreat is indeed discovered... at last. And that is our journey's end' (10.192). 'Our purpose', as the narrator declares, 'is to track the rolling wheels, and bear the travellers company on their cold, bleak journey' (10.197). Letters may provide knowledge of Nell and her grandfather, which knowing both represents and enables a connectedness inseparable from the letter's mobility, but not until the single gentleman meets up with them, dead and deluded though they will be, will he have reached a point from which he can retrospectively assemble the two halves of this story paradoxically about their paths not coming together.

As his post chaise races on, the hitherto 'unfathomable' (9.111) single gentleman recounts in a longish, dense speech the history revealing Nell's grandfather as his brother that explains his pursuit of them: 'I have a short narrative on my lips...' (10.197). In accounting for his life, however, the single gentleman does not, as one might reasonably have expected, lay out some lengthy personal history pointing to all that he has been doing in the meantime. On the contrary, he has tellingly little to say about himself. He explains that he is the younger brother of Nell's grandfather, that as a child he was sickly and his older brother did much for him, and that when as young adults they both fell in love with the same girl, the single gentleman, loving his older brother deeply and feeling indebted to him, 'left his brother to be happy... [and] quitted the country, hoping to die abroad' (10.198). So does the single gentleman's story literally recount his purposing his own heterosexual matrimonial singleness. And that is not all. A more generalized, unaccompanied solitariness becomes virtually everything the reader need know about the rest of his life: 'The younger brother had been a traveller in many countries, and had made his pilgrimage through life alone' (10.199). This efficiently disposes of practically an entire lifetime of activity: having occurred 'abroad' to a 'traveller', whatever happened can be understood to evaporate into an anonymous outland of internationally disconnected action. Readers need not worry about what this character has been doing in the meantime. He has not been in that time zone. Instead Dickens hones the single gentleman's personal story to his bare trajectory toward his reuniting with his brother. His appellation sums him up. This character embodies overcoming his 'singleness'.

As Dickens clarified in a memorandum summarizing the single gentleman's speech, its purpose is to provide a scaffold enframing the whole tale as his. First, the single gentleman pushes back the tale's beginning. After unveiling himself as the grandfather's brother and beloved childhood companion and accounting for their parting long ago, the single gentleman briefly relates a knotty story of marriages, children, and deaths in his brother's family whose effect is to bring the narrative rapidly down to the grandfather and Nell at the beginning of the tale, now pegged to himself. 'So the old man and child in opening of story', Dickens annotates (memorandum). Having thereby sketched in the personal history of the grandfather after he, the single gentleman, had departed (pieced together from rare letters), the single gentleman skips forward to tell in a few words how he decided to pack up and 'with the utmost speed he could exert' rejoin his brother, and how he subsequently 'arrived one evening at his brother's door' (10.199) in London only to find that his 'Brother and [the] child have disappeared' (memorandum). The germane cotangent to this arrival too late is that Nell and her grandfather have not vanished—narratologically—by leaving the country, as he did. This story relates their (pedestrian) disappearance and the single gentleman's (coach) pursuits. In a final annotation, Dickens wrote: 'So the single gentleman when he first pursued.' He thus condensed the single gentleman's actively pursuing his brother within the story and the single gentleman doing so (in passive tense) pursued by the story's narrator.

The single gentleman flying along in the coach thus provides a retrospect that re-plots the story within his own unrealized journey to rejoin his brother. This retrospective narrative does not simply set up the single gentleman as yet another character with a personal past en route to encounter Nell and her grandfather. The story of his story is the story. It includes Nell and his brother's divided-off journey, which, not illogically here, is connected to him in being disconnected from him. As Dickens's memorandum makes explicit, the single gentleman in his coach—the single gentleman and his coach—present a potential line through an individual that connects the whole narrative...or rather, this wholly broken narrative....

The omniscient narrator has been providing this glue, and, in the first half of his speech, the single gentleman signals his relation to that omniscient narration. He relays his history in an omniscient-like third-person. Here is the single gentleman: 'I have a short narrative on my lips': 'There once were two brothers, who loved each other dearly. . . . The youngest . . . had been a sickly child. . . . He left his brother . . . he quitted the country . . . The elder brother married . . .' (10.197–8).

Because readers have already been reading the centrally divided, omnisciently-told, retrospective multiplotted novel that relates the subsequent part of this story, it makes perfect sense that the single gentleman's revelation of his and his brother's unattached adult lives shifts abruptly when he reaches his London entrance into the story. As his interlocutor, Mr Garland, says: 'the rest I know' (10.199). The reader does too. The single gentleman can now quickly abridge: 'We may spare ourselves the sequel. You know the poor results of all my search. Even when, by dint of such inquiries as the utmost vigilance and sagacity could set on foot, we found they had been seen with two poor travelling showmen; and in time discovered the men themselves—and in time, the actual place of their retreat; even then, we were too late. Pray God we are not too late again!' (10.199).

Having rendered the first part of his speech in the third person, the single gentleman switches over into the second person (we, you). This too is utterly logical, if grammatically indecorous. Within the story world, where the single gentleman has just provided to Mr Garland a third-person narration of himself, the single gentleman slides from a third-person, self-referent 'he' to a communal 'we' because the repeating 'we' identifies the London community he joined subsequent to his brother and Nell's walking out of it. For both real readers as well as for Pickwick listening and Master Humphrey telling, this pronominal enallage cannot, however, be separated from the tale that includes an extra-diegetic omniscient narrator who also renders the single gentleman in the third person and who creates the perspective upon him as a member of that community—that 'we' whose intertwining simultaneous activity the narrator has been tracking. In both the real and the fictional extra-diegetic layer, the single gentleman's switch to a 'we' makes sense slightly differently than it does within the tale. The 'we' continues the story he was previously rendering, but now not omnisciently, because that omniscient narration is actually continuous with the omniscient narration that is relating the scene itself and himself (as a third-person, him). No wonder the omniscient narrator declares that 'the narrator . . . stopped' (10.199) in the only such mention in the entire tale of a 'narrator'. Separate planes of diegesis are almost bewilderingly breaking in upon each other here.

Performing a quick thought experiment zeroes in on the single gentleman's near intrusion upon the tale's omniscient narrator narrating the scene he is in. If one were to imagine enabling the single gentleman to keep on narrating his history in the third person, all that would be required would be to insert a chapter break immediately after the single gentleman concludes his third-person self-narration. This is right after he describes how the younger brother—himself—'with the utmost speed he could exert...with honourable wealth enough for both, with open heart and hand...arrived one evening at his brother's door', at the Old Curiosity Shop in London (10.199). Then the story could be followed seamlessly by flipping back to the novel's thirty-third chapter, just after Mrs Jarley's 'Be in time...', after which the chapter begins: 'As the course of this tale requires that we should become acquainted, somewhere hereabouts, with a few particulars' about Sampson Brass, the story's 'historian' will aeronautically transport readers back to 'a small dark house, once the residence of Mr Sampson Brass' with its apartments 'to let to a single gentleman' (6.277). One could, in short, return to the moment when the single gentleman enters into the story.

The point here is not to remake, mind-numbingly, Dickens as Borges. The point is that the single gentleman's speech tips off that he has been playing all along, ever since his entrance, the omniscient narratorial role, which he will shortly explicitly claim as his own. The kind of surprise readers are headed for is one that has been in store.

This is not the currently accepted view. That view, which even Jaffe leaves unchallenged, is that at his story's end Dickens suddenly 'belatedly remembered' (to quote representatively from Sylvère Monod) that he had to patch his novel back into its *Master Humphrey Clock*'s frame and that he ad-libbed disastrously (as Monod voices most vociferously but also most ably).[34] Whatever one may feel aesthetically about the surprise ending, this underestimation needs to be laid to rest.

On its face, it doesn't make sense. Even if there were not powerful internal evidence and structural narrative reasons for knowing better, it is worth recalling that all through the tale Dickens has been publishing it every single week and also monthly in the cover wrapper of *Master Humphrey's Clock*, and he has been staring at the running-head 'Master Humphrey's Clock' on every proof and published page. Together wrapper and running-head make, incidentally but relevantly, for the weirdness of having to reprint, here and there, the title 'The Old

Curiosity Shop' where it was needed to prevent it seeming as though the running-head 'Master Humphrey's Clock' titled the text and thus signaled to the reader a return to the *Clock* frame story.

Moreover, there is even some contradicting external evidence. In November of 1840, shortly after the single gentleman's introduction and with the novel only two-thirds done, Dickens visited a rehearsal of a theatrical adaptation, called *The Old Curiosity Shop; or, One Hour from Humphrey's Clock*. As Dickens wrote, he went with the idea of 'preventing [the director] making a greater atrocity than can be helped of my poor Curiosity Shop'.[35] In this play, one of the characters—Mr Garland—turns out to be Master Humphrey: 'Mr. Humphrey Garland.' This choice is somewhat amazing, given that Dickens's copious autonomasia meant that the schoolmaster, the single gentleman, and the bachelor were all still at this point in the serial run seemingly better qualified to step into the role of Master Humphrey by virtue of each being, to quote Dickens, a 'nameless actor' (11.225), capable of assuming Humphrey as a surname. (Dickens originally derived the name from the clockmaker Thomas Humphrey.) The editor of the Clarendon *Old Curiosity Shop* only safely ventures that 'if Dickens did reveal to [the play's director and the author] that Master Humphrey was an actor in his drama, he did not tell them which actor he was'.[36] But saying this is to miss that the play stages, months before the serial tale's end, Master Humphrey's resurfacing as a character in his own tale. (There is, significantly, no omniscient narrator in the play.) Dickens could have threaded his story back to Master Humphrey and its *Clock* frame in a hundred inventive ways—through, say, the portraits on Master Humphrey's walls about which he discourses at length. Nor, it is safe to say, did Dickens likely steal his novel's final plot twist from the play. Not only is the theatrical 'Mr. Humphrey Garland' a concoction to frighten away imitation rather than inspire it, but also—to return to the evidence of the novel—what he makes plain is that only the already-introduced single gentleman makes sense as this story's narrator, Master Humphrey.

Instead of nullifying interpretations that assume Dickens scrambled at his tale's end, the assumption should be that Dickens felt continual pressure throughout the development of this tale into a novel to work out how his story follows from beginning to end, including both Master Humphrey's initial first-person narration and the tale's origins in *Clock*. That pressure would have been greatest about midway through the serialization of *Shop* in September of 1840, when Dickens approached the completion of *Clock*'s first volume. Dickens had then to make up the monthly number, the sixth, that included a full-blown preface to *Master Humphrey's Clock* along with a title page for it, a dedication, and the frontispiece (previously discussed). His outlook was then specially leveraged to encompass both diegetic story-worlds in which *Clock* enframed *Shop*: 'Imagining Master Humphrey in his chimney-corner, resuming night after night, the narrative,—say of the Old Curiosity Shop' (the preface, 6, p. iii). Focused on the 'whole' out of which his novel was spinning and filling up the number itself with reminders of it for his readers, in that September monthly number after leaving Nell in deepening trouble at Mrs Jarley's ('Be in time. Be in time...'), Dickens began to weave into his story its relation to its frame.

Nor should one fall into the mistake that Dickens uses the single gentleman to patch up a problem. Dickens did not really face a problem. His story was outwardly smashingly successful and inwardly well organized, especially in its centrally divided plot. Insofar as it helps to think of Dickens at all, one would do better to imagine him thinking deeply about Master Humphrey and his story and fathoming how and why his character Master Humphrey has needed to tell his story the way that he does. This novel sees its own logic through.

One may now comprehend why the single gentleman's entrance coincided with the story's making more explicit its multiplot composition through the sequencing of retrospective reconstructions of the various characters' plots by a fantastically mobile omniscient narrator. Recall that at that mid-point in the story, the narratological reassembly process that would bring the two disconnected retrospective narratives together became conspicuously both that which within the story world the single gentleman arrived to do—he is trying to reconstruct Nell and his brother's movements in order to locate them in his present—and also, extra-diegetically, that which the omniscient narrator—that aeronautical historian—calculatedly features. In the single gentleman's speech, Dickens begins to unveil that these are not two separate things.

In the closing's revelations that the single gentleman is Master Humphrey, Dickens at once both confirms and ensures that the story's narratorial structure has been all along imbricated with the relations of mobility that it has depicted. The first order of business was to disclose that the tale is structured from the personal perspective of that main protagonist, the single gentleman, a.k.a. Master Humphrey, and that the tale Master Humphrey is telling describes his tragic failure ever to connect up with his brother and grandchild. As the last chapter explained, the compositional formal structure expressing this disconnection of paths turns on the repeated failure of its two main retrospective narratives to converge. So what Dickens must do—what it is paradoxically sensible for him as well as for Master Humphrey as storyteller to do—is to go back in narrative time and retroactively undo the first-person retrospective opening chapters in which Master Humphrey's and Nell's paths converged: 'One night I had roamed into the city, and was walking slowly on in my usual way... when I was arrested by an inquiry... and found at my elbow a pretty little girl, who begged to be directed to a certain street at a considerable distance...' (1.38). This never happened, and announcing it never happened forms part of telling, part of explaining, a tragedy about journeys never connecting.

Most of what needs to be said to explain Master Humphrey's rescinding of this urban pedestrian opening has already been said in the preceding discussion of the tale's pivot between walking in the city and walking out onto the intercity highways and the alternating plot structure that then arises to relay this story of connecting and disconnecting modes of mobility. One may now summarize that story's plot. The single gentleman, Master Humphrey, races home near the end of his life to reunite with his brother. But his brother has lost not only everything he had in a gambling addiction but also his mind, and a little girl that Master Humphrey has never met, his brother's granddaughter, is trying to rescue him. This little girl has fled with his brother on a walking journey of heroic proportions, and the single gentleman tells her story and the story of his pursuit after them in a way that reveals how that walking journey meant not only that he never met this girl or recovered his brother, but also that, even though he had made his way back to the same region of the world in which they were and joined the community of which they had previously been a part, he nonetheless remained divided from them utterly. His story thus relates his total failure—thanks to their disappearing into an exiling intercity foot-journey—ever to be linked with the life of the heroic little girl who perished trying devotedly to help his deteriorating, ruined, aged brother.

In Dickens's first attempt, subsequently deleted, at presenting his tragedy's final plot twist, he focused solely on revealing the tragedy as Master Humphrey's. That is, he focused on identifying Master Humphrey as the single gentleman and undoing his role in the tale's beginning. A canceled passage in the manuscript appears beneath a ruled, separating, ink line (after the tale's last words '... so do things pass away, like a tale that is told!'):

When Master Humphrey had finished the reading of this manuscript, and again deposited it within his clock, he returned to the table where his friends were seated, and addressed them thus:

'Forgive me, if for the greater interest and convenience of the narrative you have just heard, I opened it with a fictitious adventure of my own. I had my share in these transactions, but it was not that I feigned to have at first. The younger brother—the single gentleman—the nameless actor in this little drama—stands before you now!'

Their emotions shewed they had not expected this disclosure.

'Yes my friends,' said Master Humphrey, with a saddened but placid air. 'I am he indeed. And this is the chief sorrow of my life!'[37]

In a conventional literary sense, Dickens is tying the end of *Shop* here all the way back to the beginning of *Clock*. Now readers know at last the secret spring of Master Humphrey's character that he mysteriously withheld in the opening of *Clock*: 'what wound I sought to heal, what sorrow to forget' (1.1). 'This is the chief sorrow' of his life. At a more complex level, Dickens returns to the beginning of the tale *Personal Adventures of Master Humphrey: The Old Curiosity Shop*, with its first-person narration by Master Humphrey, to untie it. Those first three chapters are 'fictitious' because Master Humphrey has had to make up this beginning in order to introduce a retrospective story that meaningfully turns out to have no beginning. He is relating a narrative that does not convey him forward to an endpoint from which he could look back and selectively mark off a beginning to the entwining of his story with that of its protagonist Nell.

The same process of reconstructive discarding, of casting off beginnings, defines the overall relation of *Shop* to *Clock*. As the urban sketch expanded and then the heroine's foot-journey takes shape ex post facto against the first monthly numbers of *Clock* involving the Pickwickians, the story works without self-contradiction to include them as written out. Within the story, right through to its surprise ending, Master Humphrey makes increasingly meaningful the Pickwickians' sidelining and Pickwick's silent listening to his story. Putting the same thing another way, Dickens remakes the challenge he faced as an author—to weave a multiplotted, serial novel out of the miscellany magazine *Clock*—fully part of his overarching meaning by expressing the fastening power of retrospective narrative to make consequential even its having been begun with a beginning that it subsequently must slough off as a beginning, the novel itself arriving 'too late'. This is why it makes sense to say that *The Old Curiosity Shop: A Tale* is written so as to be excerptable from *Master Humphrey's Clock* and yet also that its actual excerption comes at the cost of not understanding that this bibliographic and formal disjointedness fully forms part of the tragedy it relates.

In Master Humphrey's revisionary act, one can even discern the surprising value to Dickens of having a fictional character as his story-teller. Within the story world of *Clock*, Master Humphrey has 'really' lived a life. So, like a real person, he is understood to deal retrospectively with an irrevocable past in the telling of his tale. He thus must have real reasons to tell his story as he does. This is why explaining he fabricated his story's beginning does not actually swallow up Nell and the rest as merely invented fictional characters. (By contrast, as readers know full well, Dickens is actually working in a fictional mode that truly frees him to invent or reconfigure the pasts of his imaginary characters.)

Dickens had to rewrite his first version of Master Humphrey's revelation. The problem was that Dickens's first attempt leaves the story in a formally awkward predicament. Master Humphrey declares that he is the single gentleman who has been telling this story about himself within a passage that retains the same omniscient narrator as the tale he claims to be narrating omnisciently. That is no good. The omniscient narrator occupies the wrong extra-diegetic authorial level, outflanking the tale's embedding within *Clock*. It also cuts off addressing the story's omniscient telling by Master Humphrey (already rallied from within the story on the coach ride). It thus omits the pivotal act of novelistic, authorial imagination when, at the end of Chapter 3, Master Humphrey switches into omniscient narration: 'And now that I have carried this history so far in my own character and introduced these personages to the reader, I shall for the convenience of the narrative detach myself from its further course ...' (2.90).

The revised, published, surprise ending keeps all the deleted passage's revelations, but fully attends to the story's crossover from the tale's story-world to Master Humphrey's adoption of a third-person, omniscient viewpoint to relate it. Now Dickens definitively declares the 'End of "The Old Curiosity Shop"' (11.223). He then inserts a page break and labels the next page, which importantly follows directly within both the weekly and the monthly numbers, 'Master Humphrey from his Clockside in the Chimney-corner' (11.224). Here is the much maligned and neglected revelation that Master Humphrey (now returning in the first-person) makes:

I was musing the other evening upon the characters and incidents with which I had been so long engaged; wondering how I could ever have looked forward with pleasure to the completion of my tale...when my clock struck ten. Punctual to the hour, my friends appeared.

On our last night of meeting, we had finished the story which the reader has just concluded....

I may confide to the reader now, that in connexion with this little history I had something upon my mind—something to communicate which I had all along with difficulty repressed....

[There came] a timely remark from Mr. Miles....

'I could have wished,' my friend objected; 'that we had been made acquainted with the single gentleman's name...'

'My friends,' said I...'do you remember that this story bore another title besides the one we have so often heard of late?'

...'Certainly. Personal adventures of Master Humphrey...' [Mr Miles rejoined] observing that the narrative originated in a personal adventure of my own, and that was no doubt the reason for its being thus designated.

This led me to the point at once.

'You will one and all forgive me,' I returned, 'if, for the greater convenience of the story, and for its better introduction, that adventure was fictitious. I had my share indeed—no light or trivial one—in the pages we have read, but it was not the share I feigned to have at first. The younger brother, the single gentleman, the nameless actor in this little drama, stands before you now.'

It was easy to see they had not expected this disclosure.

'Yes,' I pursued. 'I can look back upon my part in it with a calm, half-smiling pity for myself as for some other man. But I am he indeed; and now the chief sorrows of my life are yours.' (11.224–5)

This passage captures the whole surprisingly fluid relation in this novel between character and what 'we call omniscience...located...between a narratorial configuration that refuses character and the characters that it requires to define itself' (Jaffe). Having reclaimed for the entire tale the first half of the tale's double title, 'Personal Adventures of Master Humphrey', which set up his initial first-person narration, Master Humphrey directly addresses his role in telling that tale omnisciently: 'I am he' and 'I can look back upon my part in it with a calm, half-smiling pity for myself as for some other man.'

What Master Humphrey explains is not only that his unique individual story has required him to tell his tale as he has, but also that, at the same time, consistent with his omniscient narration, his tale presents a perspective taking in the community as a whole in which, as he says, he merely plays a part. It was precisely the over-specifying indexing of the tale to Master Humphrey as an individual that corrupted the canceled passage. When Dickens was too preoccupied with unveiling the story as Master Humphrey's, he presented the story as solely about that specific character. In that ending, Master Humphrey concluded: 'I am he indeed. And this is the chief sorrow of *my* life!' (emphasis added). In that first rendering, his tragedy was only to be the singular 'he' who is the single gentleman. By contrast, Dickens's revised, published ending foregrounds that Master Humphrey's projection of an omniscient perspective in which 'I am he' makes the story not about one individual, but about the larger structure of community in which that individual understands himself playing a part: 'I can look back upon my part in it with a calm, half-smiling pity for myself as for some other man. But I am he indeed; and now the chief sorrows of my life are yours.' *Yours*: not just his, but everyone's, the readers' too, and the mode of his telling it has made this so.

In *Pickwick*, an omniscient, extra-diegetic narrator simply supplied an overseeing and unifying viewpoint. It was there as if by magic. But it was not magic. An individual made it up: Dickens. In essence that is the simple truth that Dickens turns around and reveals in *Shop*'s surprise ending. In a beautifully counter-intuitive reversal, the first-person opening chapters of the tale are declared 'fictitious' against the 'real' omniscient narration for the purpose of conveying the real role that fiction's omniscient narration helps play for individuals in imagining the connected relations of humans in a tale about a fateful foot-journey 'off the grid' and the coaches racing to repair the breach. This is why Master Humphrey, also known as the single gentleman, needs to break out from this road tale: to show that he both exists within it and projects an extra-diegetic omniscient viewpoint upon it. And that is not just a fate born of his journey, but of yours and of everyone's.

In thinking of Master Humphrey's omniscient narration, one might justifiably hesitate over whether Master Humphrey really can know all that he tells. And it is satisfying that Dickens has, in fact, more than ensured that he can. It is not only that, unlike the rest of the characters, the last chapter's epilogue assigns the single gentleman a biographer's role: 'For a long, long time, it was [the single gentleman's] chief delight to travel in the steps of the old man and the child (so far as he could trace them from her last narrative), to halt where they had halted, sympathise where they had suffered, and rejoice where they had been made glad' (11.222). As the very possibility of that excavating investigation indicates, the single gentleman already knows the path of Nell's journey. His parenthetical aside that he retraces it from 'her last narrative' refers back to Nell's relaying of her story to the schoolmaster at the end of her journey: 'She told him all' (8.53). 'The schoolmaster heard her with astonishment. "This child!" he thought—"...Have I yet to learn that the hardest and best-borne trials are those which are never chronicled in any earthly record, and are suffered every day! And should I be surprised to hear the story of this child!"' (8.53). The bachelor's subsequent letter, which makes unmistakable the uptake of Nell's story by middle-class men, then further represents the beginnings of the written circulation of Nell's half of the story. With such explanatory amplitude in place, one hardly needs to jigsaw in the explicit internal relaying of Nell's journey's first segment by Codlin and Short or its second by Mrs Jarley and company 'relat[ing]...all that they knew' (8.59–60), and one is saved completely from having to fantasize (what is not in the text) the single gentleman interviewing other potential sources: Dick Swiveller, Kit, Mrs Quilp. All readers need do is grant Master Humphrey literary license to dramatize his story, as his sympathetic investments require. The role of a third-person semi-omniscient narrator (which he—the single gentleman—performed on the coach) was something Master Humphrey already self-consciously repeatedly claimed for himself, having, for instance, declared that 'in treating of the club, I may be permitted to assume the historical style, and speak of myself in the third person' (2.75).

The literary sleuths out there will also delight to know that in that extra-special sixth monthly number in which Dickens introduces the single gentleman the narrator lets slip that the narrator's seemingly omniscient knowledge of others results from a perfectly mundane future reconstructive act. The telltale moment comes when Nell's troubles at Mrs Jarley's are mounting. Her grandfather is gambling again. She has become distracted from Miss Edwards, for whose friendship she wishes; and 'one evening, as Nell was returning from a lonely walk, she happened to pass the inn where the stage-coaches stopped, just as one drove up, and there was the beautiful girl she so well remembered, pressing forward to embrace a young child whom they were helping down from the roof' (6.274). The narrator, verbally relaxing, then grants a bit of information about Miss Edwards beyond Nell's ken: 'Well, this [young child] was her sister . . . whom she had not seen (so the story went afterwards) for five years, and to bring whom to that place on a short visit, she had been saving her poor means all that time' (6.274). The information omnisciently conveyed is here registered as from a 'story' gathered 'afterwards', while the contents of the scene—the stage-coaching reunion—go to prove, yet again, that Master Humphrey's knowing enough to relay the tale omnisciently serves to express its structuring relations of mobility.

In the critically-neglected *Clock* ending, the reader gets to re-see the tale *Shop*, which focused on the life and death of the individual little Nell, from the perspective of an individual, Master Humphrey, who has generated the tale's structuring omniscient narration. The difference between these two perspectives—on the one side, an omniscient narrator focalizing an individual life and, on the other, an individual focalizing omniscient narration—results in two different endings: one in *Shop* and the other in *Clock*.

In the closing scene of *Shop*, the omniscient narrator focalizes an individual character's perspective, and the ending suggests how time unfolds for individuals whose journeys through life give them beginning and endings. The tale fades out on a much older Kit, married and carrying on with his life. Kit frequently tells his children the 'story of good Miss Nell' (11.223), incorporating himself into it. As Audrey Jaffe discerns, 'Kit, in the novel's last moments, suggests a way of ending its preoccupation with Nell's image, of laying the past to rest' (69). In the tale's last words, the narrator describes how Kit 'sometimes took [his children] to the street where she had lived; but new improvements had altered it so much, it was not like the same. The old house had been long ago pulled down, and a fine broad road was in its place. At first he would draw with his stick a square upon the ground to show them where it used to stand. But he soon became uncertain of the spot, and could only say it was thereabouts, he thought, and that these alterations were confusing. Such are the changes which a few years bring about, and so do things pass away, like a tale that is told!' (11.223). In this closing image, Dickens offers a condensation of the relations of mobility that have structured the tale's telling. The Old Curiosity Shop gets absorbed into a broad road whose recent construction metonymically untethers this story's placedness and makes it instead into a story quickly dated. But this send-off also presents a way of closing a circle with the beginning in which closure comes from a sad, inevitable moving on, and 'The End' marks the point when, things having gone onto other things, the individual moves past the point of being able to retrieve the story's beginning.

Clock offers a competing and complementary closing scene. In this ending, which follows the final surprise plot revelation that Master Humphrey is the single gentleman who has told the tale omnisciently, the focus shifts to the omniscient narration Master Humphrey has produced, and an individual imaginatively focalizes omniscient narration.

From this omniscient viewpoint, communal relations actually have no beginnings or endings. Such an ending might thus even be said to be self-contradictory. What, then, might such a closing scene look like? How about having Master Humphrey pick up another, new story and hold it in his hands? And then having made this brilliant use—unique in all of Dickens—of his ongoing magazine's serial publication to perch right between the end of one novel and the beginning of another, how about offering a vision of time as marching on endlessly? Or rather, to pin the novel more precisely to its own time, how about offering a vision of London's St Paul's clock time embracing people's simultaneous activity?

As Master Humphrey returns the tale's manuscript to the clock case and withdraws a new one, he notices that 'the hand of [his] trusty clock pointed to twelve, and there came towards [them] upon the wind the voice of the deep and distant bell of St. Paul's as it struck the hour of midnight' (11.225). St Paul's London time merges into Master Humphrey's clock, whose own 'fame is diffused so extensively throughout the neighbourhood' that his neighbors set 'the exact time by Master Humphrey's clock', while at least one would 'sooner believe it than the sun' (1.4–5). These synchronized and synchronizing clocks provoke Master Humphrey to recall that he 'had seen [the clock of St Paul's] but a few days before, and could not help telling them of the fancy I had had about it' (11.225).

Master Humphrey then elaborates in great detail his clambering up to the clock. St Paul's clock may induce a fancy, but it is anything but ethereal. It is mechanized, housed, a made thing: 'a complicated crowd of wheels and chains in iron and brass—great, sturdy rattling engine—suggestive of breaking a finger put in here or there, and grinding the bone to powder—and these were the Clock!' (11.226). Nonetheless, this clock's 'pulse was like no other clock' (11.226). It seems unstoppable and inexorable: 'It did not mark the flight of every moment with a gentle second stroke as though it would check old Time . . . but measured it with one sledge-hammer beat, as if its business were to crush the seconds as they came trooping on' (11.226). The clock's pounding escapement is inescapable: 'hearing its regular and never-changing voice, that one deep constant note, uppermost amongst all the noise and clatter in the streets below—marking that, let the tumult rise or fall, go on or stop—let it be night or noon, tomorrow or today, this year or next—it still performed its functions with the same dull constancy, and regulated the progress of the life around' (11.226).

From out of this sense of St Paul's clock's unifying properties, 'regulat[ing] the progress of the life around', 'the fancy came upon [him] that this [clock] was London's Heart' (11.226):

It is night. Calm and unmoved amidst the scenes that darkness favours, the great heart of London throbs in its Giant breast. Wealth and beggary, vice and virtue, guilt and innocence, repletion and the direst hunger, all treading on each other and crowding together, are gathered round it. Draw but a little circle above the clustering house-tops, and you shall have within its space everything with its opposite extreme and contradiction, close beside. Where yonder feeble light is shining, a man is but this moment dead. The taper at a few yards' distance, is seen by eyes that have this instant opened on the world. . . . In that close corner . . . there are such dark crimes . . . as could hardly be told in whispers. In the handsome street, there are folks asleep who . . . have no more knowledge of these things than if they had never been, or were transacted at the remotest limits of the world. . . . Does not this Heart of London, that nothing moves, nor stops, nor quickens—that goes on the same, let what will be done—does it not express the city's character well?

The day begins to break, and soon there is the hum and noise of life. Those who have spent the night on door-steps and cold stones, crawl off to beg; they who have slept in beds, come forth to their occupation too. . . . The streets are filled with carriages. . . . The jails are full too . . . [so too workhouses, hospitals, law courts, taverns, markets]. Each of these places is a world, and has its own inhabitants; each is distinct from, and almost unconscious of the existence of any other.... So each of these thousand worlds goes on, intent upon itself until night comes again...

Heart of London, there is a moral in thy every stroke! as I look on at thy indomitable working, which neither death, nor press of life, nor grief, nor gladness out of doors will influence one jot, I seem to hear a voice within thee which sinks into my heart, bidding me, as I elbow my way among the crowd, have some thought for the meanest wretch that passes, and being a man, to turn away with scorn and pride from none that bear the human shape. (11.226–7)

Here an encompassing omniscient vision becomes a function of the coordinating convention of a synchronizing time, while the individual to whom this time remains indifferent learns from it not to be indifferent to those bound by it too. 'To all friends round St Paul's!' So goes an ancient toast, which Dickens knew.

But why, I want to demand, does St Paul's London clock only form the heart of London the city here? Why does Master Humphrey suddenly reappear solely as an urban *flâneur* elbowing his way through the city crowd? Whence the post-coaching, world-traveling single gentleman? This whole descriptive 'fancy' reads like something out of an urban sketch by 'Boz'. The Dickens of this passage is the famed novelist of London, author of the 'people of the city', as Raymond Williams famously dubbed him.[38] But Dickens the novelist was never delimited only to London. Indeed saying so would seem even to miss part of the significance of that city by containing it. Neither *The Old Curiosity Shop*, nor any of Dickens's novels, are only about London. London is always a hub.

'*Draw but a little circle*' around London's clock, suggests Master Humphrey. Inside that local, diurnal spacetime, one will find, day and night, a community, composed of individuals bound together in simultaneous time but mostly unaware of each other and as various as one can imagine. Yet, in the period in which Nell's story occurs, clocks set to St Paul's London time were, systematically, being carried by the fast-driving stage coaches heading day and night out of London, while in the railway present in which Master Humphrey tells his tale, the railways began publishing that time as 'railway' time in their timetables, beginning a process of synchronizing other towns' clocks and people's watches. Limiting St Paul's clock to the heart of the city of London, Master Humphrey and perhaps Dickens as well touch on historical and ideological limits around which *Clock*'s tragedy has taken shape and that, far from limiting it, lend its story some of its power.

As his tale knows better than himself, Master Humphrey's circle of simultaneous time is enlarging. The passenger transport network is expanding and accelerating. Its passengers and its interconnected cities are growing to include even that which Master Humphrey shrugs off as irrelevant because, as he remarks, 'transacted at the remotest limits of the world'. A later look at the passenger network would, then, not need to convey the same hopeful insights of *Pickwick*, nor express the same devastating tragic caveats of *Master Humphrey's Personal Adventures: The Old Curiosity Shop*. Given developing international connections to places previously seemingly 'remote', it would need to confront new international complications.

3

International Connections

I. Perspective

Little Dorrit is commonly referred to as Dickens's prison novel. As the opening three chapters pan across Marseilles' jail, then its quarantine in which travelers are impounded, and finally the London home-office its occupant has turned into her personal penitentiary, Dickens creates through that series of parallel entrapments a vista onto the world as a prison. He will proceed to build his story centrally around a family seemingly doomed to live out their lives in the Marshalsea debtor's prison, and then, after dramatically freeing them, confirm that 'it appeared on the whole...that this same society in which they lived, greatly resembled a superior sort of Marshalsea'.[1] As that sweeping analogy indicates, this novel elevates imprisonment into an encompassing symbolic figure, such that it comes to feel completely natural for the narrator to employ a phrase like 'the lock of this world' (2.47) or to describe the very rays of the sun as the 'bars of the prison of this lower world' (19&20.577).

Much of the best critical commentary—Lionel Trilling, J. Hillis Miller, John Lucas—explores this dominant theme and governing trope, imprisonment. It has now long been established that, as Hilary Schor summarizes, *Little Dorrit* bitterly uncovers a 'social sphere...terminally reducible to carceral metaphor'.[2] There are, and have been, of course, many other ways of reading *Little Dorrit*. But a distinction must be made between the carceral readings and the others. The carceral readings of *Little Dorrit* are not merely tracing one aspect among many in this compendious novel in an effort to gather together as complete a picture as they can of its treatment of all the many kinds of imprisonment it depicts. The readings of the carceral must confront the all-encompassing claims made within the story for the figure of imprisonment. They are forced toward its totalizing perspective. Other readings are not therefore disallowed, but only theirs must also be able to embrace the novel as a whole—or worry about the ways in which it cannot.

It thus can come as something of a shock to realize that Dickens also, and equally directly, warns in *Little Dorrit* that far from illuminating any greater forms of encaging, people actually 'shut up in prisons' subsist in 'a social condition, false even with a reference to the falsest condition outside the walls' (2.51). This statement would seem inconsistent, to say the least. And, as John Carey astutely protests in railing against the novel's apparent incoherence in this regard, how can Dickens suggest all the world is a prison when he continually and explicitly describes the Marshalsea and its devastating effects on its inmates with a force of comparative detail that works to distinguish it from, not conflate it with, the rest of the world?[3] There must be, one has to admit, either some hopeless contradictoriness (that is Carey's view) or something else.

That another perspective exists in *Little Dorrit* from which the prison looks quite different and not so imposing is easy enough to discover. Even as he pans in those opening three chapters across the parallel entrapments that help to create a vista onto the world as a prison, Dickens manifestly constructs another perspective around the figure of his characters' interconnected journeying. Here the world appears filled not with other inmates, but with 'fellow travellers'. Within each of those initial separate confinements Dickens shows the characters understand themselves in relation to their expected release, vividly distinguishes them by their comportment to each other as, in repeated phrase, they act and react on one another in their comings and goings, and kindles a completely different perspective in which all—the reader, the narrator, and the characters—alike are overlooking, and concerned with, their possibly interconnected plotting, plottable journeys. None of this is surprising or hard to detect. Rather what's striking is, as with the prison, how quickly Dickens builds the literal situation of the fellow travelers into another, central, complex, structuring perspective— one in which 'by day and night, under the sun and under the stars, climbing the dusty hills and toiling along the weary plains, journeying by land and journeying by sea, coming and going so strangely, to meet and to act and react on one another, move all we restless travellers through the pilgrimage of life' (1.20).

As will come as no surprise, I am going to champion, as it were, the journeying perspective of *Little Dorrit*. The carceral readings have engaged this aspect—but only so as to contain it, to expose the journeying as labyrinthine and dead-ending. In the face of this, I partly aim to advance a recent current of criticism that ignores the topos of the prison and has instead been excited by the novel's international settings, its European cosmopolitanism, its assumption that the territory across which the characters' everyday lives travel and crisscross with others is, as Dickens's own had fully become by this time, not at all restricted to England and the English.[4]

However, I am not going to offer a reading of the road or of the characters' international circulation that merely tries to retrace the figurations of journeying through the story so as to construct and analyze as complete a picture as possible of its treatment. To do so would be to miss the crucial point. Any reading of the novel's figuring of journeying must confront *its* all-encompassing claims, its totalizing view in which 'move all we restless travellers through the pilgrimage of life' (1.20). My contention is that in *Little Dorrit* Dickens reconfigures this totalizing viewpoint, inducting its readers into changes wrought by a passenger transport revolution that had, by the 1850s, become clearly international in scope. That system's international reach was, as this novel shows, transforming a long-standing perspective of a world constituted around 'all we restless fellow travellers' so as to include activity at a distance as simultaneous and to confront the perplexity, created by such an expansion, of plotting people's (diachronic) plots.

My starting point is that *Little Dorrit*'s first few chapters develop both the characters' imprisonment and their interconnected journeys such that they figure for the reader, narrator, and characters alike two contending perspectives. The cardinal—and confusing—fact is that each perspective is a totalizing one. Their separate propensities toward encompassing, and persuasively explaining, the workings of his era's social life form a part of what Dickens is showing about them. Currently, the critical understanding of this novel is lopsided, unbalanced by the misguided belief that the capacity to assimilate essentially all aspects of the novel to the carceral precludes, because it subsumes, the unmistakable other picture being drawn of journeying fellow travelers.

The actual relation between imprisonment and journeying in this novel—like many of the relations in this story—most strongly resembles a toggling of perception in which one seesaws between equally plausible realities. This phenomenon is helpfully exemplified by a host of reversible images, the best known of which is probably the figure of a duck–rabbit, first analyzed by Joseph Jastrow (Fig. 16). Looking at Jastrow's duck–rabbit helps make several relevant factors immediately clear. Most powerfully, the drawing presents competing perceptual possibilities that—as interested Ludwig Wittgenstein—are irresolvable in terms of truth or falsity. Who would debate whether one view contradicted the other? Or argue for the duck against the rabbit? Instead, the fluctuation between them feels almost mentally weightless and yet utterly determinative. Importantly, it is neither a figure–ground reversal (of the kind commonly illustrated by a vase that also profiles two faces), nor an optical anamorphosis (in which radically altering one's viewing position or optics suddenly reveals the otherwise unseeable). The duck–rabbit switch occurs to an object perceived continuously in perspective. Ears to beak, fur to feather, rabbit eye (looking right) into duck eye (looking left): each view replaces the other in its totality; the appearance of one entails the disappearance of the other. As Wittgenstein says, not perception, but the organization of perception changes, and that change occurs not so much because of any internal mental activity as because the relations constructing the external picture present more than one possible conceptual arrangement.

In *Little Dorrit* one can perhaps most readily consider in such terms Dickens's depiction of Tattycoram's maddening relation to the Meagles, which sends her careening to Miss Wade and then back again. How, for specific instance, is she to resolve Mr Meagles's demand that she count

Figure 16. Duck–Rabbit. *Harper's Weekly* (19 November 1892), 1114. Courtesy of The Huntington Library.

five and twenty to cool her anger at being treated so differently from her mistress Pet, the Meagles' daughter? Dickens undoubtedly wants readers to see it is a patronizing request, disgustingly so. It opens onto an utterly unfair and class-based power structure that has arbitrarily assigned to her the role of servant and to Pet, to whom she ought to be a sister, the role of being petted. Yet Mr Meagles's request to count five and twenty also undoubtedly comes of his honest convivial attempts to include Tattycoram in their biological family and springs from the fact that the Meagles have kindly adopted her from an orphanage and raised her from childhood. However much this oscillation may intentionally leave the reader wishing to find a way to transcend the gaping ideological social fissures so painfully at issue, the point is that whether hailed as Tattycoram or Harriet Beadle (the name assigned to her at the orphanage), she herself is, as Dickens indicates in a host of ways, riven between two irreconcilable viewpoints.

So fundamental is this kind of anti-dialectical perspectival truth to this novel that Dickens also carefully mocks its debasements into a self-serving relativistic logic through which anything may be pretended to depend upon the 'light' or 'point of view' in which it is seen. These are the terms that readers see being deployed spuriously by Mrs Gowan (13.393), Ferdinand Barnacle (18.557), and Mrs Clennam (19&20.585). Against this, *Little Dorrit* introduces two possible encompassing figurations each with the potential to supplant the other, each in its coexistence tempering the other's exhaustive explanatory power. So does 'Book One' open onto prisons impounding fellow-travelers, and 'Book Two' onto fellow-travelers on the road unable to shake off their imprisonment.

No contradiction actually arises, then, between whether *Little Dorrit* is Dickens's prison novel or his novel of international mobility. Neither does the relation between the two resolve by recognizing the symbolic power of the prison to absorb comings and goings into forms of imprisonment. A totalizing carceral perspective naturally reveals mobility to be an illusion; in fact it constitutes itself through showing it to be so—as nothing more than a parade on the prison grounds or stumblings in 'the multiplicity of paths in the labyrinth' (14.417). To adduce (again) only the most important instance, Dickens dramatizes that the Dorrits carry the Marshalsea with them on their travels, though only Little Dorrit and Frederick Dorrit admit the before-and-after parallels. Who that has attentively read this novel can forget the tragic moment when

Mr Dorrit calls for Little Dorrit to check 'if Bob is on the lock' at a formal dinner in Rome, his mind irrevocably gone back to his incarceration
in the Marshalsea prison (16.489)? Wealthy or poor, wherever you go, and
even when you appear to be going, there the prison is: the art of the novel
excels at comi-tragically showing repetitions recurring with meaningfully meaningless divaricatings, a maze without an exit and plenty of
doors. At that formal dinner in Rome, Mr Dorrit memorably salutes the
French Counts, Italian Marcheses, and English aristocrats: 'Ladies and
gentlemen, the duty—ha—devolves upon me of—hum—welcoming
you to the Marshalsea. Welcome to the Marshalsea!' (16.489).

Dickens plainly wants readers to hear, and we have, how Mr Dorrit's
salutation rings out across this entire novel and for everyone—menacing
even the reader with the sanity-shaking truth that the distinction
between being inside and outside the prison is collapsible. In Mr Dorrit's
case, his final breakdown represents a wrenching conclusion to his tor-
tured moments of lucid self-knowledge (6.165, 12.359), and those crucial lucid moments poignantly hint that Mr Dorrit's characteristic
stammering 'hum'-ing and 'ha'-ing is not only an insufferable throat
clearing tic that conveys his pompous assertion of social status, but also
a broken way of speaking that patches over his knowledge that his performance, this performance, any similar performance of status, is a piti-
ful survival tactic concocted by a lifer. As was mentioned before, the
novel's criticism has already brilliantly explored Mr Dorrit's imprisonment and how Dickens similarly renders up character after character in
their different Marshalseas: Fanny enslaving herself in an infernal cotil-
lion with Mrs Merdle; Miss Wade bitterly wreaking upon herself the
selfsame perversion of all expressions of love that her society perversely
wreaks upon the possibility of loving in despite of gender, class, or birth;
and even Britannia, sleeping in a chariot wheeled by fools on the serial's
cover, decaying within the novel like Rome before her, her potential
progress endlessly circumvented in government by Circumlocution
Office. In exploring the lock of this world, readers become, like latter-
day Miss Wades, perpetually paranoid, razor-sharp close-analyzers of
the ubiquitous maneuverings of social power—'dissect it!'[5]—always
coming to the same conclusions (all roads leading to Rome)—about its
inescapability, its tyrantless tyranny. One may discover only too quickly,
with or without Foucault's or D. A. Miller's help, that the panoptic
prison can be the master key to which it lays claim. Welcome to the
Marshalsea![6]

We will be departing almost immediately. I only want to pause for a moment to notice that peering through the all-shadowing bars of this prisonous world, readers and critics inevitably confront the novel's redemptive note, the undeniable evidence that Dickens does not dissolve everything into utter hopelessness, as a kind of mysterious problem. Where does it come from? How can it be sustained or justified? To take the three critics previously invoked: Lionel Trilling will scrape up from the example of Daniel Doyce, the engineer-inventor, the vague salvific possibility of discovering peace in an unselfish creative expression of will; J. Hillis Miller will grasp at the idea that childhood innocence can persist (plunging him into an unfortunate misunderstanding of Little Dorrit, who is anything but childlike); and John Lucas, having penetratingly shown how this novel forecloses the pastoral and nature as potential backdoor exits from the bonds of social imprisonment, will conclude simply that in characters like Cavalletto and Pancks and most especially in Little Dorrit, one discovers 'affirmative possibilities... existing within the city-prison' (p. 286). Lucas's is, I believe, the most common resolution. Readers of this novel are, with specious large-mindedness, required to accept an inoculating dose of irresolution, and, recognizing a vastly shrunken field for individual agency, to concede that nonetheless somehow value resides in quotidian little doings—those Little Dorrits, for whom we give Pancks, our cavaliers having been diminished into Cavallettos.

This dubious escape hatch from the carceral amounts essentially to saying that this novel proffers the muted moral counsel that one ought still to decorate one's cell with flowers and, being careful not to delude oneself too much, to swallow one's prison rations as if they were a saucy feast. This notion that Dickens pins all hope on society's inmates consoling themselves through well-intentioned self-deceptions accords especially powerfully with Little Dorrit's loving final request that Arthur Clennam burn a piece of paper (presumably his great-uncle's will's suppressed codicil) that would, it is understood, reveal to him the truth about his real mother. The novel's penultimate scene, before the quiet final wedding, thus presents the striking image of Little Dorrit hiding from Arthur the awful truth about the past that he has been striving since the story's beginning to uncover. Only, the argument goes, such acts of 'pious fraud' (2.56) imposed upon others or even upon oneself can relieve the grim social and psychological internments that are universally people's lot in life. Inevitably, if not always explicitly, the dully conclusive conclusion to this didactic and reductive line of thinking is that Dickens's novel as a novel models the enacting of willful deception, call it fancy, that its story, even as it warns against fantasy, appears to prescribe. Shall I say it once again, Welcome...?

Attempting to sustain the carceral perspective as exclusively and ubiquitously true, as it suggests itself to be, ultimately comes to feel not so much wrong or untrue—that is not the right criterion—as Cyclopeanly single-minded. The trouble, as Jastrow's duck–rabbit helps clarify, follows from a failure to discern (what anyone may daily experience to be true) that alternative totalizing perspectives can coexist, and distinctively in this case that there can be a co-possible perception of the organization of relations that does not necessarily contradict or undo the authenticity of the encompassing figure of the carceral.

It is not merely that the novel opens onto a Marseilles full of people moving through it from everywhere—'Hindoos, Russians, Chinese, Spaniards, Portuguese, Englishmen, Frenchmen, Genoese, Neapolitans, Venetians, Greeks, Turks, descendants from all the builders of Babel' (1.1). The characters Dickens introduces self-awarely relate to all this international human motion in which they take a part, and they respond to the different shapes that same awareness takes in others. This reflexive and intersubjective orienteering complicates even the opening chapter's ostensibly melodramatic opposition between the villainous Rigaud and the good Cavalletto, focusing readers upon the not so easily categorized variances in their outlooks onto the international terrain in which both explicitly situate themselves. Thus Rigaud, the selfish murdering drifter, sinisterly detaches himself from the people and environments he manipulates, declaring himself (what one might possibly be inclined to admire): 'a cosmopolitan gentleman...own[ing] no particular country...a citizen of the world' (1.7). In contrast, the generous John Baptist Cavalletto, 'a poor little contraband trader' jailed for helping undocumented 'other little people' across international borders in his boat (1.7), turns out to be not only firmly Italian but specifically from Genoa (famed historic center of shipbuilding). His first acts are, significantly, to pinpoint the time by studying his surroundings (a 'clock' Rigaud calls him) and then to sketch, as if upon a trip, a European map on the floor describing their current location (1.3–4).

Far from captives assimilated to a captivity that captures even their movements, Rigaud and Cavalletto, along with the rest of the characters introduced in the opening chapters—including, grotesquely, Mrs Clennam with her canting eyes on the afterlife—all treat in their different ways their physical imprisonments as nothing more than haltings from which to take bearings amid, and upon, their unstoppable journeying.

In the grip of a perception organized around seeing everyone in relation to each other on their journeys, the prison suddenly dematerializes as the novel's quintessential figure for describing modern life. Arthur Clennam's confinement in the Marshalsea becomes, for instance, nothing more than one among several means of experiencing a 'marked stop in the whirling wheel of life' (18.545). His perspective upon the passage of his life runs right through the prison, as the narrator explains in the chapter 'The Pupil of the Marshalsea' (18.545). And that chapter title—'The Pupil of the Marshalsea'—itself neatly exemplifies the kind of polysemy that allows for the toggling figurations. 'Pupil' has two primary meanings, denoting either a student or an aperture of the eye, and from a perspective organizing all in terms of imprisonment both meanings would apply. Just as in the story's larger sweep an encompassing carceral perspective finds self-confirmation in Clennam's eventual confinement in the prison he once visited as a patron, so too it gathers semantic micro-force here by incorporating the double-meaning of 'pupil' as a condensation of the truth that Arthur must face that he too is the disciple of a prison that holds him in its all-seeing eye. But, then again, understanding all the same happenings in terms of journeying, both meanings of pupil—educational and ocular—apply again, but differently. Arthur Clennam's imprisonment represents a final step in the antidotal narrative arcs that he and Amy each trace toward the other, and so, student-like, this hero is learning from Amy's model before him (while she herself undergoes an edifying version of his past experience), while, in visual terms, Arthur's imprisonment in the Marshalsea provides a salutary lookout for this traveler back onto their intertwined journeys. Thus does the word 'pupil' look out of this novel, reading the reader reading it—foregrounding the formation of perspective. And, if its petite instance of double framing a double-meaning were not enough to induce the perspectival light-headedness that the hero Arthur repeatedly experiences, one can even see—one may blink to see—the competing incongruous inflections coiled up all over again inside that final key word, 'Marshalsea': 'Martial see' or 'Marshal the sea', two senses resonating separately in two different gestalts.

While, then, Pickwick airily announced, 'now...the only question is, Where shall we go to next?' and sped off by stage coach, and while Nell wondered to herself 'Which way?' and trod fatally forward on foot, Arthur Clennam, crossing the Mediterranean sea en route from China, has that question posed to him by a stranger befriended on board a ship: 'And now Mr. Clennam, perhaps I may ask you whether you have yet come to a decision where to go next?' (1.15). Hedging and tentative, self-interrupting and cautious: the question of going retrenches here into one about a prior mental process of 'com[ing] to a decision' to do so. (The answer can be: 'no, I have not...', 'yes, I have...'.) Syntactically Dickens foregrounds people interacting around their thinking about their going as much as their actually going.

In fact, Dickens himself made a memorandum, sketching his story's plan on the first page of his notes for the first number, about how he would organize this novel around journeying. After jotting notes introductory of the various prisons, Dickens pointedly penned a dividing line and then proposed:

People to meet and part as travellers do, and the future connexion between them in the story, not to be now shewn to the reader but to be worked out as in life. *Try this uncertainty and <u>this not-putting of them</u> together, as a new means of interest.* Indicate and carry through this intention.

The rest of this chapter will examine in detail Dickens's working out of this plan. I hope I have already suggested, however, that this novel represents more than just a mechanical enactment of the plan's crisscrossing plots. Dickens enfolds into his story as its subject the perspective capable of holding such a plan. This is what his seemingly innocuous closing self-directive to 'Indicate...this intention' hints. It is of the greatest significance that when Dickens wrote the first number he put the first indication of his original intention into the mouth of a character: 'In our course through life we shall meet the people who are coming to meet *us*...', declares Miss Wade in a much-discussed speech to which I will return. The semi-omniscient perspective by which one projects people and their lives in terms of intersecting journeying became fully a part of this story's content, an aspect of how the reader, the narrator, and the characters come to know, or fail to know, themselves.

II. Simultaneity

Before the two jailbirds, Cavalletto and Rigaud, or the quarantined English travelers come into view, *Little Dorrit's* first five paragraphs open onto Marseilles in a stunning page-and-a-half about the forming of perspective. 'Sun and Shadow', Dickens calls the chapter. He then immediately demolishes any comfortable assumptions that title might perhaps have provoked: for instance, that the sun makes things visible and shadows do not, or even that visual perspective is solely what's at issue (rather than the sun signifying time, and 'sun and shadow', humankind's oldest clocks). On the whole I suspect that many readers, at least if my own experience is any guide, venture perplexedly through their initiation into this novel. Some may understandably even doubt, as one critic wondered, if 'Dickens himself entirely knows what the language [of the novel's opening] ... is trying for ... [i]t is so strange, its significance so difficult to grasp'.[7]

I am going to analyze how and why Dickens introduces the subject of perspective in terms inseparable from this story's figural organizing of the knowledge of happenings and relations around either imprisonment or interconnected journeying, and in particular I will be concerned with the latter. If there was any call to justify this approach, one might note that Dickens divides his story into two 'Books' and that 'Fellow Travellers', the only repeated chapter title, would have headed in dramatic parallel fashion the openings of both Books were it not that this chapter, 'Sun and Shadow', like a meaningful pointer to the problem of organizing perception itself, was wedged in first. But it is hardly necessary to defend that Dickens's plan in his memorandum for 'People to meet and part as travellers do ...' on the one side and that plan's working out through the characters on the other shapes this novel's opening's obvious concern with establishing perspective. Rather the challenge is to figure out why this strange opening's setting up of a perspective upon 'fellow travellers' (as well as that of imprisonment) makes sense and what Dickens accomplishes through it.

What, I want to ask, was at stake when, with ten novels written, every single one of which opens onto England and the English, Dickens decided to open this novel internationally onto Marseilles? And why should that international opening read so strangely? The answer I will put forward concerns the totalizing nature of the journeying perspective. For without at all undoing a global and timeless-sounding claim to embrace 'all we restless travellers [moving] through the pilgrimage of life', Dickens synthesizes in *Little Dorrit* contemporary changes in imagining simultaneity across international time and space. As Dickens comes to see, even in an *English* novel 'fellow travellers' circulate in concert internationally. And, what's more, he is right. A synchronized, 'universal' standard time was extending with the international reach of the passenger transport system: not only St Paul's London 'railway time' is coming here (again), but also now the clock in Greenwich. But this is to skip ahead.

Here is how *Little Dorrit* commences—with a single line set as its own paragraph, the most compact beginning in all of Dickens's novels:

Thirty years ago, Marseilles lay burning in the sun, one day.

Insofar as this opening line intentionally echoes conventions of storytelling—'One day, at a certain time in a certain place...'—what is immediately made apparent in the four paragraphs that follow (consistent with the first line's unique typographical isolation) is that Dickens is carefully disrupting and holding up the sort of beginning in which looking back glides right into narrating forward. In fact, one could conspicuously skate over the next twenty troublesome sentences by pleating this opening line to the final line at the end of the fifth paragraph, where it reiterates ('Marseilles...lay broiling in the sun one day'). How easy it would be, Dickens seems to shrug, just to begin by zeroing in on the characters' plot lines: 'Thirty years ago, Marseilles lay burning in the sun, one day...[four paragraphs]...In Marseilles that day there was a villainous prison. In one of its chambers...were two men.'

Instead, with the opening sentence's expected 'One day...' awkwardly displaced to the line's end, where it pinches off forward progression rather than helping commence it, Dickens immediately ensnares readers in a little three-word trap: 'Thirty years ago'. These words ostensibly locate a beginning for the story in a specific past moment. Yet their real effect is not to settle the time of the story's action but to conjure a now from which thirty years have elapsed. Moreover, this introduction of an unspecified narratorial present rubs against readers' inevitable awareness of their actual, differing present, dating from which one must relate to the narrator's diegetic present time (as in, say, 'thirty years before one hundred and fifty years ago...').

Juggling these unpinioned temporalities one quickly finds that the narrator has not really transitioned into the story's past beginning, a once upon a time, but devilishly launched readers into a limbo of unresolved awareness that the past stands in relation to the present perspective from which it is viewed. No matter that by the novel's end it will be ascertainable that the story has taken place from 1826 to 1828, roughly confirming the narrator's alignment with the first serial readers' present, circa 1856–7. The backward focus announced by 'Thirty years ago' inverts that old joke (and *locus classicus* of discussions of deixis) that Dick Swiveller plays when he leaves a note on his office door, 'return in an hour' (8.80). Like *'Tis Sixty Years Since*, it provokes the question, What *is* the retrospective temporal perspective introduced?

Had 'one day' been placed at the sentence's beginning, it might have relieved some of this tension by operating to select with gusto a single definite day. Launching with 'One day, thirty years ago...' indicates a lead into defining the narrator's perspective. Transposed vexingly into a tailing clause, 'one day' generates precisely the opposite result. Shorn of its focusing narratorial function and made redundant by the sentence's less obvious but absolutely crucial time-marker, 'the sun', 'one day' threatens to dissolve the specificity it is supposed to have—languidly bespeaking instead one day among innumerable others. This unfocusing effect relies partly on the fact that nothing particular is yet really happening that might indicate the narrator's reasons for settling on this one day. Marseilles lays inertly burning in the sun, and the sentence that immediately follows, beginning the second paragraph, underscores the non-differentiation actually being evoked: 'A blazing sun upon a fierce August day was no greater rarity in southern France then, than at any other time, before or since.'

In short, Dickens conjures up the unique day upon which his story begins, a day which cuts both a past and future from itself (a 'before' and 'since'), and yet, incredibly, he prolongs nothing but the fact of the day, that day is happening, such that beyond juxtaposing the human-scale durations of thirty years and a day, he withholds any explanatory focusing of the narratorial perspective that has bothered to recollect this day and makes the reader feel—by suspending it—the absolute importance of that future to the anterior contents of the narrative. One might object that this represents a definite point of view, not its absence. The narrator's temporal perspective is indecisive, temporizing, directionless—facing the past passively. It resembles that initially of the story's protagonist, Arthur Clennam (who is also significantly neither at his life's dawn nor its dusk). But this changes nothing; the essential effect is the same. The opening defuses the narrator's conventional focusing retrospective temporal perspective, stymieing by disengaging that which it normally ostensibly enables and means: a selective attention bringing the reader into the flow of past events.

This initial thwarting of the narrator's perspective frames the more lengthy description that follows in which Dickens baffles the visual specificity he has introduced through naming the location Marseilles. Marseilles? It's there, but for most readers the most striking aspect of this novel's opening is its vertiginous confounding of Marseilles as a possible ground for visual perspective. This disruption is largely achieved through an extended play on 'staring'. Forms of staring tie together all four paragraphs after the opening line: 'Every thing in Marseilles, and about Marseilles, had stared at the fervid sky, and been stared at in return, until a staring habit had become universal there. Strangers were stared out of countenance by staring white houses, staring white walls, staring white streets...a sea too intensely blue to be looked at, and a sky of purple, set with one great flaming jewel of fire. The universal stare made the eyes ache....Far away the staring roads, deep in dust, stared from the hillside....Everything that lived or grew, was oppressed by the glare....Blinds, shutters...all closed and drawn to keep out the stare....'

The first thing to notice about this description is that the lines preceding it make it read as coextensive with a withholding of a focusing narratorial perspective. Otherwise—imagine reading it aloud—debuting with 'Every thing in Marseilles, and about Marseilles, had stared at the fervid sky, and been stared at in return...' would actually assertively pinpoint an implied past moment in Marseilles, however typical or strange, that was particularized by a blazing sun making things difficult to see. (That is how London is mired in fog and mud in the opening scene of *Bleak House*, however sweeping its symbolism may be.) Instead, as it's written, everything in scorching Marseilles may be staring, but Dickens's describing of it as such further stalls the story upon the happening of the day itself—straightforwardly baring by way of its absence how the narratorial perspective is going to form around the plotting of specific past events, now further shown to frame even the ostensibly passively receptive act of visual perception.

Readers have generally seen clearly that Dickens's opening picturing of Marseilles lacks a stabilizing visual frame or situated viewpoint. It is not difficult to recognize that seeing has become so roving and generalized as to make for a view of vision's failure; one observes that a looking lit from everywhere results in a crisis of irradiated optics in which the sun up-ends the human perspective (in its visual, spatial dimension) assimilated and literally evolved to it. But, as I just tried to hint by purposefully analyzing through visual terminology ('see', 'recognize', 'view', and 'observe'), to frame Dickens's confounding of visual perspective, when the characters do come into 'focus', this opening has also set up readers for the novel's extended, deliberate challenge to the synthesis of pictorial perspective to a whole metaphorics of knowledge and conscious awareness.[8] Sight lines turn out to be impossible without plot lines, not to mention that all real eyes are constantly on the move. When, five hundred and fifty-five pages and eighteen months of serialization after this opening, Arthur Clennam will at last realize that 'Looking back upon his own poor story, [Little Dorrit] was its vanishing-point' and that 'Everything in its perspective led to her innocent figure' (18.555), the novel that began by scattering visual perspective will have indicated not only the power of such a pictorial encoding of mental perspective, but also, even more, its inadequacy. Arthur immediately absorbs the picture into the ongoing temporal process of selectively assembling it—the story figured by journeying: 'He had travelled thousands of miles towards it...' (18.555).

Dickens is not opening with a 'picture' of Marseilles. His plan is to plot 'people to meet and part as travellers do...the future connexion between them in the story, not to be now shewn to the reader but to be worked out'. This plan commits him to an initial, imaginary, non-arbitrary, backward act of temporal selection, and understanding that helps greatly to account for the opening's narratorial filibuster as blocking the consolidation of the story's retrospective perspective. The opening can thus be seen to turn partly on Dickens's aim not merely to enact this plot scheme, but to render the kind of knowing that projects such a multistranded plot for his characters as part of their own and his reader's awareness. It is a narrative act of perception for the narrator and characters to view, semi-omnisciently, their diverging and converging courses through life. By crafting the opening as he does Dickens dismantles any assumption that such a perspective arises spontaneously or naturally or automatically as if it were somehow uncreated.

Instead, by perplexingly alighting on some one day thirty years ago for who knows why in a place that overwhelms the possibility of distinguishing anything in it, the overwhelming effect Dickens achieves is to make readers wait. This initial wait, it should be clear, is not the kind of waiting ordinarily induced by plot. For the first few paragraphs of *Little Dorrit* readers are not yet caught up in that 'curious present that we know to be past in relation to a future we know to be already in place, already in wait for us to reach it' (to quote Peter Brooks's smart definition).[9] At best in the opening readers are awkwardly only partly there. Mostly, the reader is being made to wait upon plot. It is, as it were, a wait upon a wait. Not a suspension or interruption of plot, typical to plotting: this opening is uniquely positioned as an opening to be levered in as if prior to plot. Afterwards, the characters, first the two men in jail and eventually Little Dorrit, will emerge from the light and shadow. (Literally: Phiz's first illustration pictures the men in jail, and his image silently garners some added meaning from representing the story's first visually stable moment.) And when that 'real' beginning begins, Dickens has geared readers to understand this story as about the perception of spatiotemporal perceiving, tutoring them that a past time and place does not automatically throw up a perspective enabling us to perceive that which once occurred there.

This special wait, this being alive to the anticipation of activity, is, I think, the chief technical achievement of Dickens's extraordinary opening description of Marseilles. It engenders much of the opening's strangeness by stretching the formal limits of description, and it does double work preparing for both a totalizing perspective of imprisonment (we live in a condition of endless detainment) and of journeying (we are waiting to go). As all novel readers know, description, almost by rough definition, ordinarily suspends action. This is not to deny that description advances plot or that it readily conveys all kinds of possible temporal relations embedded in that which it describes. Rather it is to designate loosely that in novels description most characteristically comes as pauses interlarded with the dialogue and with the recounting of events. Hence, though inevitably conveying the activity of perception, description nonetheless tends to represent almost, picture-like, as if it were relaying a temporally frozen field, a slice taken out of time. And for the most part, Dickens's description of Marseilles seems descriptive in just this way. For instance, in the lines omitted above in my quotation of Marseilles's blinding staring, the narrator quite naturally interweaves a whole variety of other senses besides the visual being suffocated or surfeited by the sun: its searing makes the harbor boats 'too hot to touch'; that hyperthermic temperature also prevents people from moving and thus quiets the sounds of the city; and the ovenlike heat overcooking Marseilles makes it a place 'strongly smelt and tasted'. All this, even though it concerns the swamping of perception, registers descriptively, as scene setting.

Yet Dickens is not quite producing in this extraordinary opening a moment arrested descriptively in an interim of time. Time is, in fact, continuing to pass as the narrator enters into and pans haplessly around Marseilles. Dickens observes: 'The only things to be seen not fixedly staring and glaring were the vines drooping under their load of grapes. These did occasionally wink a little, as the hot air barely moved their faint leaves.' One may wonder, Is there then some faint air current? Some movement? Dickens instead continues to sketch the scene by making absent the invisible action of the wind which had been hardly perceptible: 'There was no wind to make a ripple on the foul water within the harbour, or on the beautiful sea without.' And a bit later he adds, in a description that almost characterizes the tone of the entire opening itself, 'something quivered in the atmosphere as if the air itself were panting'.

In Marseilles at the beginning of *Little Dorrit*, time is flapping on the mast.[10] Horses, people, carts, all living and moving things, are all almost immobilized in the heat, excepting the cicadas, who eerily only appear suddenly at intervals greater than a dozen years, and the lizards, those tiny dinosaur holdovers. Formally, were he not also defeating visual perspective, the effect Dickens is creating might be compared to looking at a series of photographic stills and finding that one can barely perceive that they are actually streaming videos. 'Marseilles *lay* burning': the verb sustains the performance of quiescence where 'was burning' would make for a more purely passive past syntax. Precisely because Dickens is conveying a scene of listless standstill, hot day tarrying, he is able to blur the temporal pause in story action typical to description into the story's action.

The opening of *Little Dorrit* presents not merely aspects of Marseilles that, descriptively, seem to have existed synchronously at some past instant. It also delivers the reader, deprived of any focalizing perspective, into the past unfolding of simultaneous happenings around Marseilles. To put positively, then, what I previously registered only negatively as Dickens's retrospective introduction of a day without any indication of the events providing the principle of its selection and his coextensive stupefaction of the perceived environment, one could say that Dickens momentarily deadlocks spatiotemporal perspective in a physics of ongoing happenings that arrests any propensity toward determining causation. The passage's keyword is 'staring' because it names both action and its perception, because, like the wind, it loses the reader in their ricocheting relation. In a sense, Dickens opens this novel with a view into a world in which every downward thrust can be reframed in terms of an upward resistance, where, to put it in apposite terms of contemporary discoveries in thermodynamics, steam engines run on cold as much as they do on combusted heat, and where light by itself does not make things visible any more than shadow makes things invisible, sight requiring the two—chiaroscuro—in combination. As a result a universal, almost Newtonian, absolute spacetime vaguely emerges, and the sun takes the place of plot and makes plottedness, happenings as a mark of selection, impossible. It shines out its priority to the human spacetime whose perception it has enabled (giving a glimpse of how humans also understand that spacetime would exist without themselves).

Still—this is not Thomas Hardy—the scale remains resolutely human. The scene awaits the individual characters who will trigger the novel's as yet unassembled spatiotemporal perspective. It can feel almost too obvious to say this—to say that the unspoken burden of the reader's waiting in the opening falls upon the absent individual human protagonists, upon the people who will project a perspective in which they reappear to themselves and the reader collectively as either prisoners of self-and-society or fellow travelers on their interconnected paths: 'People to meet and part…', 'In Marseilles that day there was a villainous prison. In one of its chambers…were *two men*…', 'all we restless *travellers*'. The reason, however, to take note of their absence—those missing human triggers—is that doing so suddenly, even somewhat astonishingly, reveals the revision of conventional structures of meaning that Dickens gets by opening this novel as he does onto Marseilles. A 'French Town?' '*Yes*' (Dickens certifies in his number plan).

In the opening line, 'Marseilles' initially seems to hold out the promise of fixing the reader's perspective by fastening the narrative's point of view upon a specific nation and culture, naming a recognizable French city as the story's location in exclusionary opposition to others. It will come as little surprise that Dickens makes short work of that potential normative ordering of perspective. Instead of attaching the reader restrictively to a specific national culture, this Marseilles quickly turns out to be a potentially inclusive part of an array of national stories. This French city hosts a fully international cohort of possible protagonists. Recall the 'Hindoos, Russians, Chinese, Spaniards, Portuguese, Englishmen, Frenchmen, Genoese, Neapolitans, Venetians, Greeks, Turks, descendants from all the builders of Babel, come to trade at Marseilles'. Dickens flattens the English and with it the English language (Babel) to the middle of this pack, sixth in a list of twelve, as if to say, six of one, a half dozen of the other. He depicts Marseilles's story, like London's, or Rome's, or Venice's, as fundamentally international.

When the English novel that is *Little Dorrit* opens, however much Dickens's readers might be expecting to see English protagonists, there are—Dickens is forcing us to recognize—nonetheless virtually a globe full of interconnected humans who might step through the as yet unentered door Dickens is holding open. Who knows but that when this story first comes into focus in Marseilles it will discover to the reader that the individuals relevant to it are neither French nor English? As it happens, they will not be. (One is Italian, the other, the 'citizen of the world', was born in Belgium to a Swiss father and a French–English mother.) Here one day, this opening is declaring, the lives of France's and other nations' citizens are unfolding in Marseilles, and, just as the history of what happened in that place that one day is thus necessarily constituted by them all, so the individual lives whose stories the reader is waiting to hear, including those who are English, are on that one day coinciding and crisscrossed with the others in Marseilles.

The opening is, just as Dickens's plan calls for, looking back to the synchronous moment of time and place in which our characters' (diachronic) trajectories crossed—'People to meet and part as travellers do and the future connexion . . . not now to be shewn.' It is just that for the first few paragraphs the narrative has not yet found the characters in their simultaneity. But when it does, it will definitely not be the case that what happened this one day thirty years ago in a town abroad, Marseilles, had a later impact on the lives of the unintroduced English protagonists passing through it, and the narrator has returned to note it. On the contrary, the narrator sees that what is happening one past day in Marseilles matters immediately to those lives—forms part of what it means to retell the story of those lives. This is the grand international gesture of the opening of *Little Dorrit*. Who knows but when this story comes into focus upon Marseilles, and quite naturally discovers that the individuals relevant to it are neither French nor English, that whatever those individuals were doing on this day forms a part of the story of other, English protagonists, and vice versa? As it happens—in both senses of that phrase—it does. Here in Marseilles, two men in a jail cell, while meantime across the harbour in quarantine, some English...

Dickens thus internationalizes that perspective upon characters as 'Fellow Travellers' in which every aspect of the story's coming inter-weaving plots—its threads of diachronic action—potentially counts toward meaningfully different outcomes for the characters. What is happening synchronically in time to people in international space matters to their (diachronic) stories.

Dickens stretches meantime across an international European arena. Here, for reconfirming instance, is his introduction of the Marshalsea prison and Little Dorrit's father at the beginning of Chapter 6, in one of many echoes of the opening that echo throughout the story:

> Thirty years ago there stood, a few doors short of the church of Saint George, in the Borough of Southwark, on the left hand side of the way going south-ward, the Marshalsea Prison. It had stood there many years before, and it remained there some years afterwards; but it is gone now, and the world is none the worse without it....
>
> There had been taken to the Marshalsea Prison, long before the day when the sun shone on Marseilles and on the opening of this narrative, a debtor with whom this narrative has some concern. (2.41–2)

In *Little Dorrit*, this kind of trampolining upon international space and time is much more than merely a conventional device of omniscient story-telling reserved for the narrator to share with the reader. The whole arc of the book that carries Little Dorrit out of the prison and across Europe will work to create in her a version of the perspective of international simultaneity upon which the novel has opened. Once in Italy, Little Dorrit's basic outlook immediately becomes comparative, her present moment oddly coinciding with the daily prison routine she knows to be unfolding so far away: '[s]he could scarcely believe that the prisoners were still lingering in the close yard, that the mean rooms were still every one tenanted, and that the turnkey still stood in the Lodge letting people in and out, all just as she well knew it to be'

(11.347). And—one more, better example—here is Little Dorrit articulating to Arthur in a letter her developing consciousness of the kind of unison of international activity that the opening aims to raise into consciousness:

When we went to see the famous leaning tower at Pisa, it was a bright sunny day, and it and the buildings near it looked so old, and the earth and sky looked so young, and its shadow on the ground was so soft and retired! I could not at first think how beautiful it was, or how curious, but I thought, 'O how many times when the shadow of the wall was falling on our room, and when that weary tread of feet was going up and down the yard—O how many times this place was just as quiet and lovely as it is to-day!' It quite overpowered me.... And I have the same feeling often—often. (13.415)

Stepping back, then, to take in the strange first few paragraphs of *Little Dorrit*, it can be seen that its various special effects flow from a shaping insight that the plots of the protagonists, including those stuck in Bleeding Heart Yard where foreigners are regarded as not quite fully human, are going to be told from a perspective that convenes their individual lives as international such that all—friends, enemies, whatever their various individual relations—are 'fellow travellers' bound together in simultaneous time across their separate geopolitically demarcated national homelands. By opening onto Marseilles but having no plot begin at the beginning of *Little Dorrit* Dickens creates a moment before the act of selective attention to any plot, a space of narrative possibilities suspended in their contemporaneity ('Thirty years ago ... one day'), a 'meantime ... ' prior to any coordinated interrelations, so that when the plot does begin it is, as Dickens planned, imaginatively descending into several synchronous plot threads whose contemporaneity binds them together across an international terrain.

This envisioning of the individual plots of his characters as unfolding simultaneously in an international arena is something subtly but essentially different from, and disruptive to, the international viewpoint Dickens has held up until this point. Previously Dickens had readily and repeatedly shown that his British protagonists traveled the world and had contacts, frequently life-altering, both at home and abroad with other nations and their citizens. From Count Smorltork's preposterous misperceptions about England in *Pickwick* through the extended, transformative journey Martin Chuzzlewit takes to America, up to the recent, unusually Francophobic portrayal of the assassin Hortense in *Bleak House*, Dickens's international perspective firmly recognizes that the individual plots of his own nationally-hybrid British protagonists have sent them journeying worldwide and that that world has also inevitably and powerfully made its way to them at home, especially in London. The view is neither necessarily xenophobic nor incompatible with national-ist prejudice. The pivotal thing is that in such a view foreign contacts or foreign presences at home may have deeply shaping and defining impacts, but the structure of meaning is such that other countries and what hap-pens in them as well as the effects they have at home operate obliquely upon the British protagonists. This tributary status is what the lateral international simultaneity of *Little Dorrit*'s opening chapters is announc-ing has changed. A reconfigured internationalist view imagined around international simultaneity enters to compete with what next to it will begin to look like a persistent, older internationalist framework for look-ing upon the world to take account that what happens over there affects what will happen here, and vice versa.

In a historically materialist way, ocean-going steam ships were, espe-cially for the British Isles, the main passenger substructure enabling this transformation. It matters that, taking dates from *Little Dorrit*, in the 1820s—when the story is set—passenger steam ships were already plying their paddles between Britain and nearby nations, while in the 1850s—when Dickens was writing—a transition had begun to iron-hulled, screw-propelled, giant steam ships—the age of ocean liners. One of the most famous anecdotes of the passenger transport revolu-tion is about the ability to comprehend and embrace its international reach. In 1835, at a Directors' meeting of the Great Western Railway, the directors were nervous about building the unprecedentedly long rail line. Isambard Kingdom Brunel squashed their worries, not by addressing them, but by sweepingly declaring that they ought to extend the railway further, to the coast, to Bristol's port, and then add a steam boat to carry people to New York. They did. In 1838 Brunel's wooden-hulled PS (for paddle-steamer) *Great Western* was launched. In 1842, Dickens himself made his first Atlantic crossing on the *Britannia*, the Cunard line's first paddle-steamer, fully equipped with sails as was typ-ical. It was a miserable little boat and a miserable long trip, but the future that ship represented was dawning. The next year (1843) saw the historic launch of Brunel's iron-hulled, screw-powered SS—for steam ship—*Great Britain*, which first crossed the Atlantic in 1845.[11]

In this context, it's worth noting that, after finishing *Little Dorrit* in May of 1857, Dickens followed up a dinner in Greenwich by going to check out 'the great ship'—the largest ever—then under construction, Brunel's *Great Eastern*. There is a famous photograph of Brunel, taken sometime that same year, that shows him in front of the chains used in the ship's launch (Fig. 17). This image can serve here to capture trans-oceanic passenger transportation's outsized linking together of individuals, and it perhaps also usefully suggests an entirely off-kilter way of perceiving the chains of links connecting the letters D-o-r and r-i-t on the novel's monthly wrapper (Fig. 18).

Notice, too, the tiny steam ship pictured on *Dorrit's* wrapper, crossing on the central-lower horizon. Turn the page: a harbor, and wait; 'There was no wind...'.

Figure 17. Isambard Kingdom Brunel and the SS *Great Eastern* (1857), by Robert Howlett. © National Portrait Gallery, London.

Figure 18. *Little Dorrit*'s monthly wrapper. Courtesy of the Department of Special Collections, Young Research Library, UCLA.

Merely invoking this revolution in international passenger transport is accompanied by the potential danger of making all that has been said seem to lapse into the dumbest of arguments. One of the worst possible misunderstandings of my argument in this chapter titled 'International Connections' would be to believe that I am suggesting that a global or an international perspective *per se* is new: as if *Little Dorrit* could announce—surprise!—that England and the English were not alone in the world. The specter of such foolishness must be raised to be banished only because the rise of fast public transportation systems can precariously present as if it were the beginnings of the global circulation of people. In reality, its meaningfulness derives precisely from what it changes about that already-present global circulation. The indispensable starting point, both for Dickens and also for my own argument, is that people had crawled the world over—literally from time immemorial and literarily in all memorialized time. Internationalism is completely characteristic of the picaresque tradition of the English novel from Thomas Nashe's *The Unfortunate Traveller* (1594) through Defoe and on. To whatever extent Dickens was aware that nations had only relatively recently arisen as a dominant political form, in his thinking of Albion he never forgets, as one might do in a nationalist fantasy of island insularity, that it was once Roman. On the contrary, *Little Dorrit* is premised on it—and on the difference between then and now. In Rome himself one time, Dickens found himself annoyed by repeatedly encountering some English tourists, and he observed that 'Mr. and Mrs. Davis, and their party, had, probably, been brought from London in about nine or ten days. Eighteen hundred years ago, the Roman legions under Claudius, protested against being led into Mr. and Mrs. Davis country, urging that it lay beyond the limits of the world.'[12]

Little Dorrit renders up the changing spatiotemporal perception of 'the limits of the world'. As its opening shows, the unsettling of the relations of time and place reconfigures conventions for constructing contexts of time and place. The world has not simply gotten smaller and travel cheaper and more convenient. Altering the spatiotemporal 'limits of the world' uncovers some limits that had been silently inhering within a comprehensive perspective upon 'fellow travellers': reformulating a perspective in which, for ordinary individuals, their interconnected movements come to unfold across an international terrain together in shared, synchronized time.

What is new about their interconnectedness is that it happens in simultaneous time because there no longer seems to be the same distending, dilated experience involved in getting from here to there. Think of how, for people today, the quickness of airline travel can be more disorienting than arduously traveling the same route by car. This spacetime warp was first most strongly felt on a physically smaller, national scale in Dickens's early life by the escalating speed of the stage-coach system. Hence one finds Dickens in an 1835 sketch observing that 'the passengers who are coming in by the early coach...are evidently under the influence of that odd feeling produced by travelling, which makes the events of yesterday morning seem as if they had happened at least six months ago, and induces people to wonder with considerable gravity whether the friends and relations they took leave of, a fortnight before, have altered much since they left them'.[13] Then here is Dickens after tripping by steam to Paris in 1854 registering again just that feeling: 'that queer sensation born of quick travelling, which will poke your Garrick dinner of yesterday down a perspective of at least three weeks, and fill your head with the strangest incoherencies concerning days of the month'.[14]

As Dickens's comment upon tripping to Paris registers, international distances were not just being made shorter and easier to cross, bringing foreign cities closer together; time—motion measuring motion—was being carried across those distances and national borders by individuals circulating swiftly and continually. As a result, time itself correspondingly underwent a slight tightening reorganization around people's coordinated circulation; synchronicity was being imagined across those once long distances in a historical horological change— time's standardization—consonant with the internationalist opening of *Little Dorrit*.

As I discussed in earlier chapters, over the course of the nineteenth century all the various local times being kept across Great Britain— time modifying longitudinally—gave way before a single unified 'railway time'. 'There was even railway time observed in clocks, as if the sun itself had given in', the narrator glooms in *Dombey*.[15] This railway time was first based on St Paul's London time and then subsequently on the observatory's at Greenwich. Time's standardizing amounted within England only to an adjustment of minutes, but before its coming towns that now may seem merely a short distance apart did not necessarily keep the same time. Why should they? They not only kept their own local time historically, but also the sun would be noon in one first and in the other later, roughly matched to the moment the dial clock's hands pointed up together (noon), when the sun's ante meridiem (a.m.) became post meridiem (p.m.). By the sun's gauge, for humans it was not the same time in those two places. Each place was its own meridian. Time derived from keeping people in one area attuned to their mean solar day.

Though it is a fascinating story, there is no real need to recount here the details about this coming of 'Greenwich time for Great Britain 1825–1880', as Derek Howse has done expertly in his chapter of that name in his book on Greenwich time, and as a few other historians—surprisingly few—have done elsewhere.[16] What matters here is that Howse is typical in framing his account as about the establishment of a national time. The story, as he tells it, recounts what happened in 'Great Britain'. Yet as Howse's and every other historical account also makes perfectly clear, it was not nations or their governments who brought this simultaneous time into being for ordinary people. 'Great Britain' only belatedly sanctioned a fait accompli. In doing so, it provides the source of Howse's 1880 end-point: 1880 marks the nation's legal adoption of Greenwich time. But, that endpoint tellingly contrasts with Howse's choice of an 1825 beginning. Howse chooses 1825 as a start date so as to begin with the first passenger railway, the Stockton–Darlington.

For the individuals who experienced its coming and whose inter-locking schedules of movement it served, standardized time was mani-festly only partly about unifying their national time. Picture, for a moment, Dickens sitting in the Ship Inn in Dover. He is pausing there for a few days' rest at the end of April 1856 on his way back from visit-ing France, struggling to start the eighth number of *Little Dorrit*. He is writing instead a little sketch for *Household Words* about how easy it is to be distracted from writing and to waste time, and in this moment that is not particularly special or unusual, he relates among other things (that is his point, there are so many other things!): 'I had scarcely ... dipped my pen in the ink, when I found the clock upon the pier—a red-faced clock with a white rim—importuning me in a highly vexatious manner to consult my watch, and see how I was off for Greenwich time. Having no intention of making a voyage or taking an obser-vation, I had not the least need of Greenwich time, and could have put up with watering-place time as a sufficiently accurate article.'[17]

First here, there is the topicality of Greenwich time that raises it to a level worth mentioning for Dickens. Where in the 1840s clocks had begun appearing with two different minute hands, one keeping local and the other railway time, in the 1850s local time capitulated to the establishment of Greenwich time across Great Britain. An oft-repeated claim is that '[b]y 1855, 98% of the public clocks were set to GMT'.[18] It is also commonly observed that in 1851 many people encountered this railway time in taking the train to the Great Exhi-bition in London. In 1855, the year before Dickens writes the sketch, across the Downs and up the coast in the town of Deal (where in the sketch he subsequently walks off to), a time-ball had been erected that dropped by electric pulse direct from Greenwich. Dover would follow with a Greenwich time gun. This telegraphic electrification was not establishing Greenwich time. It was automating an already widespread distribution of Greenwich time that was being sustained by manually carrying clocks. For this reason it has to be said that Dickens's comment is a little askew, a bit behind the times, 1840s-ish. His 'watering-place time' already is Greenwich time—isn't that after all what the pier clock is importuning? And an 1852 map showing which towns were on Greenwich and which were still keeping to their local time confirms as much (Fig. 19). How could it have been otherwise? The South Eastern Railway, which Dickens has been riding en route to Paris via Folkestone since 1850, had come up to Dover years before, in 1844, and that was two years after the Railway Clearing House recalibrated the separate company railway times to Greenwich.

One could at this point telescope backward through the process of standardizing time that brought Dickens to notice this pier clock declaring Greenwich mean time. Its recent history extends to John Palmer's 1784 reform of the Royal Mail system around stage coaches that carried a clock in a locked box set to London time, propagating time through the transport system. Its deeper history reaches back through the eighteenth-century spread of watches and the rise of individual time disciplining, whose import to literary history (especially in relation to the rise of the daily diaries both private and published) has been superbly told by Stuart Sherman in *Telling Time*.[19] And its further origins carry on back through the invention of pendulum clocks (inventor: Christiaan Huygens), which standardized the length of hours, previously varying with the seasons, along with the astronomical advances that led to the founding of the Greenwich Observatory. We are circa the mid 1600s, but one can easily push still further into the past—into, say, the Middle Ages, to Richard of Wallingford's grand invention of an automated 'celestial theater', horological harbinger of a world running on mechanical clocks.[20]

All of that history and much more (I artificially constrained it to England) is relevant, which is why I mention it here, to that pier clock and to the international simultaneous time onto which *Little Dorrit* opens. Yet, at the same time, Dickens's remark also helps to sort out more narrowly how Greenwich time relates to his era and to *Little Dorrit* and to the nineteenth-century passenger transport revolution.

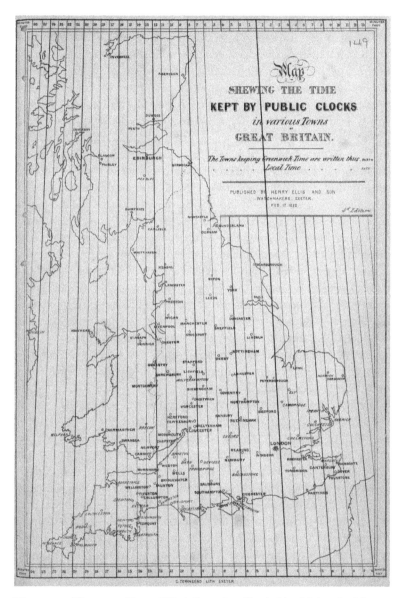

Figure 19. Time map. Henry Ellis & Son, 1852. Cambridge University Library, MS.RGO.6/597, fo. 149r. Reproduced by kind permission of the Syndics of Cambridge University Library.

What Dickens says is: 'Having no intention of making a voyage or taking an observation, I had not the least need of Greenwich time.' Dickens not being an astronomer, that last bit about taking an observation is just a bit of fun. It includes only to wave away the scientific, cosmological dimension of Greenwich mean time. By contrast, however, pausing for a few days' rest on his way to London from France, that is, in the midst of a voyage, Dickens is actually straightforwardly invoking the fact that he can momentarily dispense with Greenwich mean time. That time is for individuals traveling. It is for people going places, fellow travelers on the move. And that is in fact Dickens's reason for producing this detail about Greenwich time in this sketch. Sitting idly in his room, distractedly looking out his window, he is defining his current state against that clock time, thereby concisely and unlaboredly conveying how his ordinary self-directing purposeful movement through his environment has been overtaken by whatever momentary distraction the environment happens to present. That general predicament is what this sketch—neatly epitomizing an aspect of the sketch form—is really sketching. The clock on the pier collapses from sustaining the temporal field in which he had been moving to become just another object catching his eye. The whole sketch, called 'Out of the Season', is narrated from just this temporality that is felt to be an oddly enjoyable but disconcerting dropping out of time. (Going off the grid need not always be tragic.)

Is this Greenwich time, a national time? Partly it is, but primarily for Dickens it signifies voyage time. And, this is the reason why as part of this discussion of *Little Dorrit*'s opening it helps to reconstruct this image of Dickens looking out of his window in the Ship Inn at Dover at a pier clock announcing Greenwich time—an English traveler from France, a pier, Ship Inn, Dover, all helping make real the point that from his individual viewpoint the standardization of time Dickens experienced extended with the reach of the stations and piers of a transportation system and the circulation of individuals in it. 'Terrestrial Time' and 'Cosmopolitan Time': these were names proposed later in the nineteenth century for the worldwide clock synchronization now officially called Coordinated Universal Time.[21]

The revolution in passenger transportation may have abetted the production of national time, generative of simultaneity within borders, but, as any expert in locomotion could see, the Greenwich time on the clock on that pier was not restrictively national. By the mid-1850s, just as it was obvious that the passenger transportation revolution did not stop at Dover, England's ancient gateway to the world, but, on the contrary, swept on across Europe and was in a process of hooking up network by network across the globe, so too the world thus interconnected was becoming united temporally in a new way as well as newly disconnected. (In *Little Dorrit*, for instance, the narrator notes the characters keep 'English time' in Rome where time-keeping was still ecclesiastically tied to post-sunset prayers, 14.446.)

A century after the introduction of Palmer's Royal Mail coaches, twenty-five nations met in Washington DC in 1884 largely to iron out the wrinkles that they themselves as nations caused in the standardization of time. The vote there was to establish a common prime meridian from which to fix longitude and time. Scientific necessity mandated having it lie with a world-class observatory, but equally important and yet impossible was that it be a politically neutral selection. It needed to transcend nations. Notice, in this regard, how on the 1852 map of public clocks (Fig. 19), the artist represented, oddly but logically, Greenwich time as a line that extends right through the title information as if it overlaid, or transcended, its English label and almost somehow the map's publication itself. The best thing to be done turned out to be the most practical thing, and what most persuaded conference delegates was Sandford Fleming's table showing Greenwich's existing worldwide predominance for ships currently operating. Though Fleming had in his mind's eye the ships' navigators, his reasoning goes right to the heart of the international composition of standard time around the coordinated movement of individuals: 'if...the convenience of the greatest number alone should predominate, there can be no difficulty in a choice'.[22]

On the one side, then, an international passenger transport system and on the other: not nations, but individuals, the people of diverse nationalities, who were its riders. And in between: the adoption of Greenwich time subtly redefining the keeping of time.

The older mode of keeping time helped hold an epistemology, a structure of meaning, in place for human activity. It meant that anything done at a distance from home matters only when it comes across distance, which is in fact time, to affect home and in doing so necessarily occupies a position in the past, however tremendously important its effects might be. By contrast, standard time is premised upon the semiomniscient awareness that what is going on is going on simultaneously at different places, such that the relation—the shrunken distance, two trains speeding toward each other—between these places now amounts to a zone of human contact, a space shared in time. (In *Little Dorrit*, China marks the far-flung realm of ongoing, separate human activity that is still as yet too far off to exist co-laterally.) The brain-bending fact for modern readers, since we live inside the reality of standardized time, is that two clocks at a distance synchronized to Greenwich time, even if they are arranged into different time zones, do not record simultaneity. They produce it. Simultaneity, like time itself, is a product of convention.

Isn't this change a key to why one might begin a story by fore-grounding the sun—that imprecise old clock—only to show that sun shows that nothing but day is happening, and then follow this opening up by having that sun give way to the movement of individuals—characterized as 'fellow travellers'—whose synchronous terrestrial action unfolding and interconnecting diachronically across an international arena (albeit one far from evenly interconnected) turns out to be a basis for both meaning and time, which are—as ever—inextricable from each other? Showing the simultaneous happenings *within* the place of Marseilles only required the local time of the sun moving overhead. But the fact that Dickens shows that simultaneity matters to an international set of characters, that it at once engulfs and depends upon the traveling of the English and other protagonists, means that readers come to look upon the Marseilles of the opening, as Little Dorrit will later look upon the tower of Pisa, as a place of simultaneous happenings *within* the diachronic story of those protagonists.[23]

If the coming of standard time had really been only about establishing a single national time for Great Britain, the nation as one meridian, under which all clocks chimed together, then it would be perfectly accurate to tie it to its creation for individuals, as it surely does partly create for them, a sense of national unity in their simultaneity. But it was not. As *Little Dorrit*'s 'fellow travellers' makes clear, individuals are meaningfully connected across international space in simultaneous time. Or rather, Dickens, shaping the novel as a form specially built around crosscutting plots in 'meantime', helps bend the universe slightly to make this real. One may suddenly see coming *A Tale of Two Cities* (transfiguring the history of the historical novel), then *Great Expectations* (exploding the *Bildungsroman* form), and finally *Edwin*

Drood (collapsing the distancing of a 'mysterious' Orient). None of those markedly international novels nor *Little Dorrit* makes any sense if one clings to the notion that the novel's form is one in which its characters acting 'meanwhile' simultaneously works to sustain the imagined community of the nation. That old insular pod snaps (Podsnap!).

As ought to be made completely explicit at this point, I am offering a reconsideration of Benedict Anderson's brilliant, ground-breaking argument in *Imagined Communities: Reflections on the Origin and Spread of Nationalism* (1983). As Anderson saw: 'The idea of a sociological organism moving calendrically through homogenous, empty time is a precise analogue of the idea of the nation, which also is conceived as a solid community moving steadily down (or up) history. An American will never meet, or even know the names of more than a handful of his 240,000,000-odd fellow-Americans. He has no idea of what they are up to at any one time. But he has complete confidence in their steady, anonymous, simultaneous activity.'[24]

The sustaining 'Meanwhile...' correlation between novel and nation that Anderson explains fully applies, I believe, to Dickens's novels through *Bleak House*. But Dickens's later tales, developing his insights into the public transport revolution, help expose an unsustainable slippage in Anderson's argument that the meanwhile structure of the novel enables the imagining of national community.

Once one begins to consider that the all-important simultaneous 'homogenous, empty time', which Anderson argues enables these nation-states, does not actually stop at national boundaries, but on the contrary indicates, and partly springs from, an interlocking global passenger network, from railroads and route maps, steam ships and time-tables, then one may also begin to see that this time is not, as Anderson assumes, something actually empty to be filled in by a national community. It becomes part of the contents of a system that binds people together as coordinations in time and space, as contemporaries.

Partly at stake in grasping this distinction is how the literary history of the period gets reconstructed. In the grip of an axiom that makes the simultaneous activity of individual citizens the bearers of national culture, one might suggest the history of nineteenth-century fiction from Scott to Dickens charts the return of an outward-looking ethnographic gaze upon Empire back upon its center, coming not so much from those foreign peripheries, but emanating from the center itself: an auto-ethnography, as James Buzard has suggested. And one would discover, as Buzard does, that this perspective reaches its zenith in Dickens in *Bleak House* (confirming why it is useful to read *Little Dorrit* as a response to that novel). Similarly, one might argue for overturning a long tradition of framing English national literary history with London at the center and the peripheries on the periphery by demonstrating how that center follows the nation-building lead of its peripheries. Thus do Ian Duncan and Katie Trumpener call attention to the Scottish novel and the Scottish antiquarians' resurrection of their bards, while Priya Joshi, along with others, has highlighted that it was first in India that an English national literary curriculum was established.[25] Of the many literary histories self-consciously concerned with the forging of empire and nation, Ian Baucom's *Out of Place* in particular expertly reaches right up to and into the limits of the struggle to draw national historical cultural identities for individuals from locales, confronting (instead of choosing between) all the contradictions and bewilderments of the 'identity-endowing properties of place'.[26] One might mark the difference between Baucom's approach and the one taken in this book by observing that Baucom takes the hybrid Gothic-orientalist architecture of the Victoria Railway Terminus in Bombay (now Chhatrapati Shivaji Terminus in Mumbai) as one of his touchstones, picturing it on his book's cover as well as insightfully interpreting it within. In contrast, this book focuses on a competing formulation that belongs to the people of all nationalities paying little heed to that architecture and instead rushing through the station to catch their trains.

Yet how does—one might object—an English novel written in British English render up international simultaneity? This question goes right to the crux of Anderson's other main argument in *Imagined Communities*. Anderson explains also that the novel abets the consolidation of nations around specific vernacular national print-languages, and this half of his argument has also rightly become widely familiar and influential. Arriving in a world already filled with languages, but where print-capitalism is converging with print-technology (mass-producing newspapers and serial novels), nations reproduce themselves partly through 'the development of a standardized language-of-state', Englishness finding a unifying cultural basis in the English language even in the face of the failure of language and nation to map isomorphically, one-nation-to-one-language: 'These fellow-readers, to whom they were connected through print, formed...the embryo of the nationally imagined community.'[27] To imagine an international community in terms of fellow travelers, Dickens would, by contrast, need to break through any such linguistic coalescing around English, coextensive though it might be with his own novel's articulation in English.

Right from the opening, Dickens directly confronts this delimitation of his English novel to the English language as at stake in picturing international simultaneity—Marseilles contains 'descendants from all the builders of Babel' (1.1). After all: who knows but that when this English novel first comes into focus in Marseilles it will discover to the reader that the individuals relevant to it, who are neither French nor English and whose actions that day form part of the story of our English protagonists, may not even be speaking English? Altro! As it happens (I write in English), even though one is Italian, they are all speaking French. Despite—or just because—this linguistic twist is first directly confirmed in a parenthetical authorial aside—'(they all spoke in French, but the little man was an Italian)', 1.4—one may recognize that this thread of the internationalism of the story is not a trivial detail, incidental to the fact that it conducts readers through places where languages other than English are spoken. Instead one may recognize how in *Little Dorrit* Dickens is repeatedly and powerfully bringing his story to the brink of its own linguistic delimitation as written for English language readers—that for Dickens to open onto Marseilles writing in English, he means partly to confront this inwrought tension of the novel's national form.

'No more of yesterday's howling, over yonder, to-day, sir; is there?' (1.11). This is how Mr Meagles opens Chapter 2, which introduces the English protagonists in the quarantine and enunciates the novel's first 'untranslated' English words of dialogue after a whole chapter's worth the reader understands to have been conducted in French. Meagles is reporting, without knowing it, that he has heard the dismayed reaction to the release of the villain Rigaud from the jail across the harbor. This sound without language is all that as yet connects the two sets of characters whose lives will all closely cross in not just one but two plot lines later, but for Meagles the howling represents his acoustic *disconnection* from the people of Marseilles. From this remark, he will go on to lightly disparage the French, and specifically the people of Marseilles, associating them with their eponymous famous Revolutionary song, national anthem of the French Republic, a people always regressively 'allonging

and marshonging to something or other' (1.11). The irony, which
Dickens is setting up, is that Meagles would have been howling right
alongside them at Rigaud's release; that he is distancing himself in
nationalistic terms from 'these people', as he repeatedly calls them, who
are, as Arthur Clennam gently replies (five lines under the banner of
that chapter title 'Fellow Travellers'), just acting like 'most people' (1.11);
that, in short, he separates himself off from what he hears going on
around him in Marseilles, when in fact readers know that what is happening meantime—inside that howl—forms somehow part of the
larger story in which he too plays his part. It is perhaps not too much
to say that Dickens thus shows that the capacity to apprehend the international simultaneity he is creating through the two separate chapters
of parallel action in Marseilles depends upon the capacity to hear foreign speech as something more than a howling—even when it is a
howl. Failing that, one ends up with the insular internationalism of
Meagles, 'who never by any accident acquired any knowledge whatever
of the language of any country into which he travelled' (1.17), and who
instead blunders about Europe right through till the end of the story:
'Still, with an unshaken confidence that the English tongue was somehow the mother tongue of the whole world, only the people were too
stupid to know it, Mr Meagles harangued innkeepers in the most volu-
ble manner... [a lengthy comic description of his performance in this
vein]...and was in the most ignominious manner escorted to steam-
boats and public carriages, to be got rid of, talking all the while, like a
cheerful and fluent Briton as he was' (19&20.610–11).

What Dickens is indicating here is not merely that in quarantine in
Marseilles or at dinner in the Saint Bernard Monastery or elsewhere,
the English characters should, as they generally do, participate in conversations conducted in French or other languages, as they wend their
way through a multilingual Europe. This novel's structure of international simultaneity requires the reader first to dive in among an international set of characters speaking only in French in Chapter 1, and
then repeats the same complete displacement of the English characters
and English language eleven chapters later when it next picks up
Rigaud and Cavalletto upon their accidentally meeting again in
Chalons as they travel their separate routes toward England. Each self-
contained chapter, unfolding the journeying of two foreigners as part
of the plot, represents a puncturing jolt into another arena of simultaneous human action that is being wholly conducted in a language

other than English. In this context, Meagles stands, at some level, for the danger of the English novel's own silent mode of proceeding confidently as if 'the English tongue was somehow the mother tongue of the whole world'.

So: an opening in English onto Marseilles, followed by an account in English of a scene in a jail cell conducted in French, and then a leap across the port to some English characters in quarantine and specifically to Meagles: the nadir of language as a barrier.

The novel goes on to array some jokes around mis-encounters between English and foreign languages. Conversing with a Frenchman in quarantine with Pet translating, Meagles subsequently mistakenly hears the speech of a foreign language as if it were actually English. He pounces on the Frenchman's 'Plait-il?' (beg pardon?), given in the text in French, taking it to be—the reader must silently deduce—the assertion 'quite ill' to which he replies insanely: 'You are right. My opinion' (1.17). An expression of linguistic incomprehension (beg pardon?) thus neatly gets mistaken by Meagles as a corroboration of his views. Where Meagles hears French as English, elsewhere Dickens plays on a Frenchman's ear's capacity to hear foreign words lurking in English. Most readers likely catch that Merdle alludes to *merde* in French, which in a classic Dickens comic deflation makes shit 'the name of the age' (12.362), but only those listening carefully will also realize that it is likely this homonymic resonance, and not the supposed awesomeness of Merdle, that explains why the otherwise highly competent French Courier—speaking English but hearing in his head French—has 'agitation in his voice' when he is faced with announcing Merdle and instead botches and super-anglicizes its enunciation: ' "Miss' Mairdale!" ' (15.462).

Some further important comic wordplay affirms the connective power of bilingualism. When Cavalletto is run down by a fast-moving mail coach, as Arthur Clennam immediately understands upon coming upon him (though the reader has heard nothing from Cavalletto), he is crying out for water. 'First, he wants some water', says Clennam (4.117). Oddly, however, before Clennam's arrival, the 'general remark going round, in reply' among those who also speculate that the injured man is 'Porteghee' or 'Dutchman' or 'Prooshan', has been 'Ah, poor fellow, he says he'll never get over it; and no wonder!' (4.117). The only way to make sense of this is to hear the crowd's hearing in English Cavalletto's crying for water in Italian and French (again Dickens does not give it): 'Abbeverare! L'eau!' Heard untranslated as English, these words together sound like 'I be very low.' This particular little multilingual joke nicely redresses an exchange in *Nicholas Nickleby* in which Nicholas, asked the name of water in French, says, '*L'Eau.*' Lillyvick, the straight man, responds 'Lo, eh? I don't think anything of that language—nothing at all.'[28] Where in *Nickleby* Lillyvick is simply made to look foolish by Nicholas, who self-complacently notes all languages are equally useful in having names for everything (as if translation were a process of word substitution) and who signals nothing much more with his French than his social standing within England, by contrast in *Little Dorrit* the power of understanding a foreign language is conveyed through the fact that Clennam's connection to Cavalletto is made to depend on it. Most importantly, his ability to fathom partially his own past and present situation stems from him and Cavalletto eventually discovering their mutual knowledge of Rigaud through a French song rendered partly in English, partly in French: 'Who passes by this road so late? Comapagnon de la Majolaine!...'.

The narrator of *Little Dorrit* never notes, much less explains, these language games. Instead, at the risk of his readers missing them, Dickens grants a silence that potentially trains the reader to listen, and think, with a multilingual ear in contrast to the characters who are the jokes' butts. And this is key. Readers are mildly asked to overcome the linguistic disconnection.

Dickens's opening in English onto Marseilles, where French is spoken, turns out, then, to set up silently for an extended reflection on the polyglot reality surrounding both his novel (addressed to English-language readers) and his characters (who can overcome the language barrier). *Little Dorrit* almost reads as if Dickens had thought to himself that a 'first principle of this book is that the meaning of the "nationalism" of national literature, and more generally national culture, cannot be understood from within the context of that single national—here, British—culture.... [T]he history of culture...needs to be written across national cultures and...it is fundamentally not a single narrative but many, [and it includes]...the nationalization of language.'[29] That quotation, however, is not Dickens, but Irene Tucker—outlining her aim in *A Probable State* (whose theoretical critique of Anderson galvanizes my explication here).

As a central play in this effort, as every reader of *Little Dorrit* knows, when the characters talk a foreign language, Dickens writes his English in such a way as to indicate the foreignness of their speech. 'To the devil with this Brigand of a Sun', 'Say what the hour is', 'How goes the world this forenoon', 'Death of my life!'—all examples of 'French' from Chapter 1. If Dickens had not been brilliantly joking with language so much, it would seem only too easy to make a joke of these encounters between English and a foreign language, to laugh at these attempts, which seem vaguely reminiscent of the dialogue of an Elizabethan play, to indicate the presence of foreign speech through a combination of stiltedly reproducing literally the syntactical constructions and denotive meanings of other European languages, faithfully copying their idiomatic phrases, and sometimes even interspersing or referencing a foreign word or two (in macaronic mode).

And, in fact, Dickens has directly outflanked any attempt to parody his narrator's English renderings of a foreign language. Her name is Mrs Plornish. And Mrs Plornish, standing shoulder to shoulder with Mr Meagles, pointedly helps to clarify further how Dickens renders the English language as both a connective force and a barrier in this novel. Recall that Meagles, who travels the world, fails to recognize that the implication of his refusal to hear or speak any other languages besides English is also a failure to recognize the actual synchronous relevance of that international terrain to himself and to his life, which relevance the novel is working to present. In contrast, Mrs Plornish, who never goes anywhere, turns her own failure to speak a foreign language (Italian) and her substitution of a kind of wacky version of English for it (in caricature of the narrator's performance) into a means of recognizing and participating in the international arena in which it turns out she must, like the novel, locate her English-monoglot self. So, while one is certainly meant to laugh at the fact that ' "Me ope you leg well soon"...was considered in the Yard but a very short remove indeed from speaking Italian' (7.223), nonetheless, as that example also makes clear, the form and content of padrona Plornish's speech establishes—and is premised upon—her instantiating her beneficial connection to Cavalletto. Performing as a kind of faux translator confirms her real relations with Cavalletto, in contrast to Meagles who chokes off his real international relations linguistically.

Mrs Plornish thus also marks the nadir of English language as a barrier in a completely different sense of nadir. Having seen in Meagles how the novel's linguistic delimitation to English could create a barrier that blocks the recognition of the international interconnectedness in simultaneity, one sees in Mrs Plornish that the overcoming of this barrier amounts to much, much less than actual facility in other tongues—and this for the obvious reason that the actual connections between people are not determinedly and only linguistic. If the novel's treating the English language as if it were 'somehow the mother tongue of the whole world' is patently absurd, potentially fatal for meaningful international human interaction, nonetheless Cavalletto is on his way to the hospital when Arthur meets him, the crowd literally carrying him there is outraged at the speeding mail coaches that have injured him, and after his recovery a single, polysemous, expressive Italian word, 'altro' (as in 'Hallo, old chap! Altro!' 'Altro, signore, altro, altro, altro!'), will, as Dickens prominently dramatizes, suffice for a bond of friendship between himself and Pancks (7.224). Dickens thereby makes sharing a single word that itself denotes 'other'—altro—into the diglot bridge of international dialogue, 'altro' all the Italian one needs to know to know Cavalletto, whose navigation of his course through life speaks louder...

Just because the Marseilles pictured in the opening in English is a place where French is mostly and officially spoken it is not therefore indescribable as a place of simultaneous action for the English or anyone else. Rather the international perspective that forces the novel to confront its English-speaking existence in a multilingual world also reveals language is only one possible way of connecting people, just as the novel knows itself as one medium among others. (Hence a universalism can readily coexist along with the fact that the novel's address to English-language readers produces Anderson's linguistically-connected national community. This ambidextrous effect is completely familiar: the novel as a genre is always rediscovering itself as both thoroughly national and nonetheless addressed to all humankind.)

And while, as Dickens shows, there are countless ways that people are connected—economic, familial, romantic—the umbrella figuration (besides imprisonment) for perceiving people as linked together internationally in *Little Dorrit* is as 'fellow travellers'. Passenger networks here subtend discourse networks. So when, as the title of Chapter 3 announces, the lead protagonist arrives back at his London 'Home' after a twenty-year absence in China, that home has already been dramatically reframed by an international synchronizing of 'Fellow Travellers'.

Trying to understand himself and his past, Arthur Clennam will look over his old London home for answers that he believes are hidden there somewhere. But meantime a man is on the move—the reader has barely glimpsed him before he cynically drank the 'ancient civic toast' to 'all friends round St. Paul's' (1.32)[30] and then set out abroad. This man is carrying a locked box, filled with secrets, away from that London house to Antwerp. There he will meet a drifting murderer coming from a jail cell in Marseilles, Rigaud. The (diachronic) plot of this novel has begun, and it too has something new to say about fellow travelers.

III. Plottability

Dickens, one may safely guess, would not have been much surprised at the results of Stanley Milgram's 1967 experiment indicating that people ostensibly remote from each other were typically linked by no more than six degrees of separation. Dickens got 'the small-world problem', as it is called.[31] 'The world', John Forster reports Dickens often remarked, 'was so much smaller than we thought it; we were all so connected... without knowing it; people supposed to be far apart were so constantly elbowing each other.'[32] That experience of a 'small world' relates in obvious ways to the passenger transport revolution, and, as Dickens helps to show, it involves not only ploddingly connecting the dots of plottable trajectories. It involves a whole problem of knowledge and mobility since, for instance, the very course of the characters' trajectories depends on the limited and fragmented amount each currently knows or does not know about their interconnections 'in their coming and going so strangely, to meet and to act and react on one another' (1.20).

My argument here is that *Little Dorrit* represents how individuals within this system, who project an omniscient-like view of it, nonetheless must perpetually confront that their own perspectives are always partial, incomplete, and belated. Their difficulty is not—as some later modernist writers will have it—that their modern world is fragmented and disconnected. It is just the opposite. The density and extensivity of people's interconnections exceeds their capacity to grasp them, producing a bewildering incompleteness in the understanding people have of what's going on around them. This was a lesson born of Dickens's viewing the passenger transport system internationally, but it applied equally within the nation and ultimately addressed the abstraction of the system's systematicity.

Little Dorrit's plot, insofar as it figures crisscrossing journeys, holds the key here. Critics, who generally otherwise prize this novel's tightly sculpted complexity, have commonly deemed its plot hopelessly snarled. Sometimes they assert that Dickens simply fails to pull the threads together as he promises; sometimes they assert that the plot makes sense but that it feels arbitrary and forced; and, most persuasively, sometimes it is suggested that Dickens is here purposely defeating plot. In all these views, *Little Dorrit's* plotting, more or less explicitly, at best reconfirms its prison-world in which 'all any of us can do is circumlocute'. In its sharpest treatment as such, Sylvia Manning has argued that the plot's failure is fully intended and represents a subversion of the novel's form: 'the plot notion of significant journey has been parodied from the start'.[33]

This apparent failure of *Little Dorrit's* plot has appeared most clearly rooted in Miss Wade's dramatic prophecy in Chapter 2, 'Fellow Travellers'. After Mr Meagles and Pet bid farewell to Miss Wade in quarantine—'Good bye! We may never meet again', says Meagles—Miss Wade darkly rejoins: 'In our course through life we shall meet the people who are coming to meet *us*, from many strange places and by many strange roads . . . and what it is set to us to do to them, and what it is set to them to do to us, will all be done' (1.19).

As has been broadly understood, Miss Wade's declaration proffers a figure for this novel's plot. She indicates the story's journeying plot structure, originally outlined in the number plan ('People to meet and part as travellers do . . . '). Readers subsequently discover, however, that Miss Wade has previously been romantically involved with Pet's suitor, Henry Gowan. Her omniscient-like prognostication—echoed more softly by the narrator at the chapter's end—thus comes to appear completely disingenuous. Small wonder Pet detects hostility embedded in Miss Wade's supposedly detached phrasing: 'It implied that what was to be done was necessarily evil' (1.19). An omniscient eye looking down impartially on Pet and herself, both moving antlike on their respective plot trajectories toward unforeseen collisions like the present one? That is simply not the case. Miss Wade withholds that she is stalking Pet, and prophesying about what one knows and controls doesn't really count as prophecy. (I predict, by the way, that there will be more on this aspect later.)

This recognition—that the crisscrossing destiny Miss Wade augurs later gets undercut—forms the kernel of a whole scholarly line of thought, whose sharpest articulation and exploration is, as I mentioned, by Sylvia Manning. 'The hints about roads of life converging do not pan out', Manning believes, and this is proof of Dickens's bleak message in which 'the plot notion of significant journey has been parodied from the start'. Recently, Hilary Schor has usefully begun chipping away at this view that, 'isn't this plot, quite simply, a failure?' As Schor protests, the novel 'works quite diligently to make *us* connect the various parts of the book...leading us to ask just how determined these plots are...How much, to return to Miss Wade's earliest statement of narrative determinism, was "set" to be done to us, and by whom?'[34]

In pursuing such a partisan plot summary, the first thing to remember is that Miss Wade is first and foremost a 'self-tormentor'. Her dark declaration actually makes much more sense if it is understood to be principally directed primarily toward her own self and only indiscriminately and generally inclusively aimed toward Pet. It is telling that Miss Wade will never, not even after Pet's marriage, reveal Henry Gowan's philandering, though it almost certainly began after he had begun courting Pet. Cowering closer to her daddy—'O, Father!'—Pet 'in her spoilt way', a phrase Dickens added in his careful revisions, 'childishly' misunderstands Miss Wade (1.19).

In line with Miss Wade's characteristic destructive self-analysis, Miss Wade is dissecting the state of her own affairs even as she looks upon another's, and her declaration to Pet that 'In our course through life we shall meet the people who are coming to meet *us*...' reads, reframed by Miss Wade's viewpoint—and not Pet's—as Miss Wade's resentful philosophizing about her own experience and about the scene in which she currently acts. From Miss Wade's perspective, she has in fact come to see who has randomly crossed her path, and, in usurping her lover, done to *herself* what was to be done. As Miss Wade later explains to Clennam: 'It was not very long before I found that he [Henry Gowan] was courting his present wife, and that she had been taken away to be out of his reach...I was restlessly curious to look at her...I travelled a little: travelled until I found myself in her society, and in yours' (16.507). So while Miss Wade and Pet Meagles in quarantine have not been 'shaken out of destiny's dice-box into the company' of each other like the two prisoners across the bay, Cavalletto and Rigaud (1.7, repeats 3.96), or 'thrown together by chance' (1.17), as Mr Meagles says in his toast to the assembled travelers in quarantine and as he and Clennam have been, Miss Wade's omniscient-like view of the intersecting journeying of who knows who still applies fairly enough to her own self. As Dickens later reveals, Miss Wade has simply generated this semi-omniscient perspective from her own experience.

Moreover, Miss Wade pointedly continues her speech. Almost no one bothers to quote her speech's second part because, I can only assume, it appears that Miss Wade merely restates what she has just said.[35] She forecasts: 'you may be sure that there are men and women already on their road, who have their business to do with *you*, and who will do it. Of a certainty they will do it. They may be coming hundreds, thousands, of miles over the sea there; they may be close at hand now; they may be coming, for anything you know, or anything you can do to prevent it, from the vilest sweepings of this very town' (1.19). The larger significance of her trans-oceanic, international elongation of the road ('hundreds, thousands, of miles over the sea') has been discussed. The question to ask here is what exactly Miss Wade is insinuating by suggesting that the people coming might be 'close at hand'? That suggestion poses an immediate problem for readers who have just read about two jailbirds across the harbor. Is Miss Wade knowingly referring here to the prisoners just 'over yonder' introduced in the first chapter? Is she already somehow acquainted with Rigaud, that 'low, mercenary wretch', as she will later repeatedly call him, and is she referring to him in referring to people coming from—Dickens added the adjective in revision—'the vilest sweepings of this very town'? When, one needs to calculate, did Miss Wade meet the novel's central villain, Rigaud?

It's a crucial plot question, and it goes to the heart of the international journeying plots to reconstruct that Miss Wade has no idea, nor will she ever, that the mercenary wretch she later meets in Italy was a murderer rotting in prison across the harbor when she confronted Pet in Marseilles and made her oracular declaration.[36]

What happens? How do the journeys intersect? In order to comprehend this plot that is figured as converging journeys, one needs to grasp that, while this novel's overall plot has many branches, it is also fundamentally split between two separate strands. Let's call the novel's secondary plot the 'romantic plot'. This plot concerns Miss Wade, who had an affair with Henry Gowan before the story begins. The novel's primary plot concerns the Clennam and the Dorrit families. Let's call this plot the 'locked-box plot'. It turns on a locked box that the servant Affery sees Flintwinch's twin brother spiriting away (after drinking a toast to 'All friends round St. Paul's') from Mrs Clennam's London home. The papers in this locked box reveal, as Rigaud the blackmailer will say, that 'There are the devil's own secrets in some families' (9.267), and, leaving aside for the moment some intricacies, the family skeleton here is that Mrs Clennam is not the protagonist Arthur Clennam's biological mother.

My concern here is not primarily with sorting the details of these two plots, but with the way in which they get related to each other. Unraveling the two plots in terms of their journeying structure involves reconstructing the path of the novel's villain Rigaud as he grifts his way around Europe. The first hint of this comes in Chapter 11, 'Let Loose', when the narrator suddenly shifts away from the multiplot activity building in London to pick up with 'one man, slowly moving on towards Chalons' by the river Saone in France. It is Rigaud on his road, 'a very long road', 'a cursed road' (3.91). Now calling himself 'Lagnier', Rigaud stops at the Break of Day Inn, where he runs into Cavalletto. Both these two foreigners to whom the reader was introduced in the first chapter are, the reader now learns, not only out of jail but also separately making their way to London. The brief chapter closes with the narratorial eye pulling back to an omniscient, journeying view of Cavalletto fleeing from Rigaud at the actual break of day: 'When the sun had raised his full disc above the flat line of the horizon, and was striking fire out of the long muddy vista of paved road . . . a black speck moved along the road and splashed among the flaming pools of rain-water, which black speck was John Baptist Cavalletto running away from his patron [Rigaud]' (3.96). Here they come, the reader may think. And sure enough, back in London, the narrator will subsequently muse that it would be:

Strange, if [Mrs Clennam's] little sick-room fire were in effect a beacon fire, summoning some one, and that the most unlikely some one in the world to the spot that *must* be come to. Strange, if the little sick-room light were in effect a watch-light, burning in that place every night until the appointed event should be watched out! Which of the vast multitude of travellers, under the sun and the stars, climbing the dusty hills and toiling along the weary plains, journeying by land and journeying by sea, coming and going so strangely, to meet and to act and re-act on one another, which of the host may, with no suspicion of the journey's end, be travelling surely hither? Time shall show us . . . only Time shall show us whither each traveller is bound. (5.129–30)

At first glance, the narrator's soliloquy may seem like an anticipatory gesture. It reminds readers, and especially Dickens's serial readers

('Time shall show us...'), that possibly Cavalletto, and most likely the traveling stranger Rigaud, is on his way. And yet—this is key—time does not actually show that Rigaud is on his way to join the plot in London. Time shows Rigaud is joining that plot in Antwerp ('in the Cabaret of the Three Billiard Tables, in the little street of the high roofs, by the wharf at Antwerp!', 19&20.591)—right about now. There, as readers will only understand much later, Rigaud is meeting Flintwinch's twin brother (Ephraim) and learning the secret contents of the locked box. The previous findings of this chapter apply: to understand this novel's (diachronic) plot, one must retrospectively reconstruct a synchronous international terrain. The narrator is thus not, one may belatedly realize, clairvoyantly foreshadowing the future. The narrator's predictions erupt to tell of things to come because of plot events that are, as yet unbeknownst to the reader and the characters in London, happening in Antwerp.

Hence when Rigaud appears on Mrs Clennam's doorstep, he is not joining the plot in London. He is already part of it, having joined it in Belgium. He initially misrecognizes Jeremiah Flintwinch—'Death of my soul!... Why, how did you get here?' (9.257)—because, as his confusion betrays, Rigaud has met Jeremiah's twin Ephraim, glimpsed momentarily by the reader hundreds of pages earlier: 'You are so like a friend of mine! Not so identically the same as I supposed...' (9.258), Rigaud remarks, having come from Ephraim. Rigaud presents Flintwinch with a letter of introduction ostensibly from the Paris correspondents of the House of Clennam and Co: 'We have to present to you... M. Blandois, of this city... &c. &c.' (9.259). But it is all a lie. It is not the contents of this letter that count, but the hand that it is written in, Flintwinch's brother Ephraim's: 'No doubt you are well acquainted with the writing. Perhaps the letter speaks for itself...' (9.259). From here on, through the rest of the chapter, clues to which the reader does not have a clue, all to do with the contents of the locked box, proliferate. Rigaud's conversation is palpably laden with incomprehensible innuendoes: he banters with Mrs Clennam about the watch containing the monogram D.N.F, 'Do Not Forget'; he mocks a portrait of Arthur's father as part of a happy family; 'Holy Blue! There are the devil's own secrets in some families, Mr. Flintwinch!' (9.267).

Having retrospectively understood that Rigaud discovered the contents of the locked box and is reconnoitering the potential for blackmail in Book One, one can then begin to understand the significance of the fact that when Book Two opens onto the Alps, Rigaud, now alias 'Blandois', has drifted into the company of Gowan and Pet on their honeymoon trip. '*Rigaud to meet Gowan and his wife*', Dickens inscribed at the top of his memorandum for the number. Miss Wade (stalking by proxy) will subsequently hire him because 'her curiosity and her chagrins awaken [her] fancy to be acquainted with their movements, to know the manner of their life' (18.568). As even Rigaud will not realize for some time, he has now become enmeshed in two separate plots—the romance plot as well as the locked-box plot—that both involve a number of overlapping characters.

By having Rigaud reappear with Gowan and Pet in the Alps and then having Miss Wade hire him in Venice to spy on them for her, Dickens structures this novel such that the main characters are so densely interconnected that they find themselves interacting in not just one plot but two. As Dickens explained to Forster: 'In Miss Wade I had an idea, which I thought a new one, of making the introduced story so fit into surroundings impossible of separation from the main story, as to make the blood of the book circulate through both.'[37] Figure 20 shows a map reconstructing Rigaud's path.

In Book One, Rigaud, en route from Marseilles to London, discovers the locked box in Antwerp and then in London scouts the potential for blackmailing Mrs Clennam and Flintwinch. In Book Two, he reappears with Pet and Henry Gowan in the Swiss Alps, and after his travels with them in Italy, he returns to London. Now, however, he is enmeshed in two separate intrigues with an overlapping cast of characters.

The relation of this novel's two central plot strands structures the story as about fellow travelers failing to be able to know or track those plots, whatever might happen to be their content. So, when, in Book Two, one night in the Strand, Arthur Clennam finally claps eyes on Rigaud walking with Tattycoram and secretly follows the two of them, he witnesses Rigaud meeting with Miss Wade ('Go on with your report...') and arranging to be paid for his clandestine services in Italy (13.399). This is the romance plot in action. A few days later (in the next chapter), a mystified Arthur Clennam, on his way to visit his old home, finds himself 'jostled to the wall' in the street by the same stranger he has just seen with Miss Wade. Rigaud is paying a visit to Mrs Clennam and Flintwinch to demand extortion for keeping the secret of the locked box (13.407). Arthur and Rigaud make their

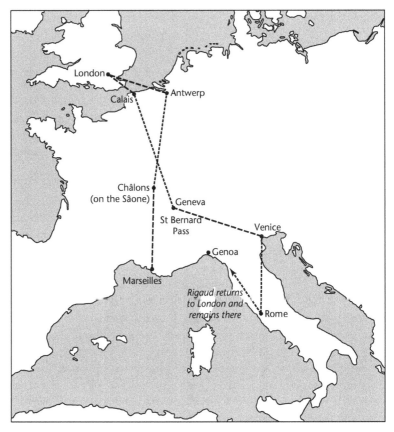

Figure 20. Rigaud's journey. His path is drawn more strongly for the portions that the story renders more strongly. For clarity, a direct path is shown from Antwerp to London. It is unclear whether Rigaud travels via Calais.

entrance together, and before being asked to leave, Arthur serves as the befuddled, Oedipal butt of Rigaud's ironic double-speak: 'Arthur?... The son of my lady? I am the all-devoted of the son of my lady!' (13.408–9).

The important point is not just that there are two separate plots— romance and locked-box—that both internally present a classic Gothic structure, in which past secrets resurge. As John Frow explains in an excellent distillation of this novel's structure in terms of genre, Dickens is subjoining the Gothic novel's plot's 'double time scale which links a surface plot to a second plot buried in the past' to a 'multistranded picaresque...[which is] essentially synchronic, a juxtaposition of simultaneous narratives'. As Frow adds, 'Miss Wade formulates this double teleology early in *Little Dorrit*: "In our course through life we shall meet the people who are coming to meet *us*, from many strange places and by many strange roads...".'[38]

One might hope for some resolution to the Gothic plots, but all bets are off once those plots become subordinated to a 'multistranded picaresque'. With that neat neologism 'multistranded picaresque', Frow names, without identifying it, the literary upshot of a passenger transport revolution through which first the picaresque-seeming *Pickwick* presented simultaneity all around the system and then the picaresque-seeming *Old Curiosity Shop* revealed the 'meantime' reconstructive process it involved. What it additionally comes to mean in *Dorrit* is that everyone is also always in perpetual process of catching up with the past simultaneous activity of others. (It was exactly this reality that Master Humphrey had to stifle in relating his story omnisciently.)

Little Dorrit's serial publication helped make this baffling untimeliness real for its readers by making the serial reader's reading always firmly belated to the activity of the writer writing. This story's scenes bewilderingly unfold from a perspective that the reader does not yet have. Small wonder, as Iain Crawford figured out, that 'for much of Book I Dickens was writing ahead of his deadlines and each number is set at a time of year roughly corresponding to the date when it was written'.[39] Where in *Pickwick* Dickens linked the serial story to its readers' present and where in *Master Humphrey's Clock* he used serialization to make real the wait on reconstructing the characters' simultaneous activity, in *Little Dorrit* he actuated the writer's actual preceding of his readers in the story. Like the characters, the serial readers were thus caught in a perpetual process of catching up with understanding because that which was going on only made any sense from a perspective lying somewhere in the future. Pipe in this novel's soundtrack. Its refrain is 'Who passes by this road so late?'

Discovering that the characters were multiply interconnected—ah-ha: there were two plots!—does not dispel the disorientation. It characterizes it. Arthur is hardly alone in his confusion. As he stumbles dumbfounded out of Mrs Clennam's room, Rigaud brazenly insinuates that she and Flintwinch have reason to murder him and then suddenly disappears, making it appear as if he really has been done away with. This disappearance looks to everyone, including to many readers, all of whom know no better than the perplexed characters, like part of a Gothic plot: is Rigaud perhaps walled up somewhere in that creepy house with its strange noises? So people are supposed to wonder, and, like Mr Dorrit, they generally do. But in fact, effecting a move he can almost never resist, Dickens has simply enfolded an aspect of the shaping formal relations of this multi-stranded road novel into his tale's contents. The novel's formal organization around Rigaud's own absent, simultaneous, international intriguing now becomes the object and topic of discussion of the characters themselves. He must be present somewhere: 'where is this missing man' (15.472)? Plot his plot.

However, just as Arthur Clennam, along with the reader, has been thoroughly baffled by seeing Rigaud interacting in two separate plots, scenes of mutual incomprehension ensue between other characters also only partly aware of their dueling interconnections. For instance, in the chapter aptly titled 'Missing', Flora goes to see William Dorrit, who is soon to return to Italy. She begs him to 'promise as a gentleman that both in going back to Italy and in Italy too [he] would look for this Mr Blandois high and low and if [he] found or heard of him make him come forward' (15.470). Mr Dorrit is excited. He knows this Blandois (a.k.a. Rigaud), but only as the friend of Henry Gowan's in Italy: i.e. as part of the romance plot. In a follow-on scene that then beautifully plays the two plots directly off against each other, William Dorrit goes to see Mrs Clennam. Mrs Clennam knows she is connected to Mr Dorrit through her secret past (the locked-box plot). He has no idea. Mr Dorrit instead naturally asks Mrs Clennam and Flintwinch if they know Henry Gowan:

'Mr. Henry Gowan. You may know the name.'
  'Never heard of it'...
  'Wishing to—ha—make the narrative coherent and consecutive to him,' said Mr. Dorrit, 'may I ask—say three questions?'
  'Thirty, if you choose.' (15.472–3)

Mr Dorrit may ask a hundred thousand questions. He will still not make the 'narrative coherent and consecutive'. The problem is not that the narrative is not coherent and consecutive. The problem is that his questions are barking up the wrong plot, and, whenever they might accidentally bear upon his hidden relation to the Clennams, Mrs Clennam will simply 'put another barrier up' (15.473). Mr Dorrit, like the reader, wants to understand what's going on? 'Hold the light for him to read' (15.472)—it will only brighten that there is a muddle. Flintwinch to Mr Dorrit: 'Now, sir; shall I light you down?' 'Mr. Dorrit professed himself obliged, and went down' (15.474). *Went down*, one might say, momentarily stealing ahead to the novel's closing paragraph, 'into the roaring streets'.

In *Little Dorrit*, Dickens renders individuals proceeding on their journeys only semi-aware of their multiple interconnections to others. Nor is it some freak coincidence that these characters happen to be in overlapping plots. Retrospectively, the perspective from which one reconstructs a story, any story, will find, Dickens suggests, that the people commonly interact with only partial knowledge of their many entwinements. Throw a circle around any limited set of travelers, say, in quarantine at Marseilles or on a tourist's rest stop on the Alps. These strangers are, and will be, interlinked though only a future perspective may retroactively select out some of the crisscrossing journeys to follow.

No one, however, can ever achieve that perspective. In the final blip of misunderstanding before the two plots begin to be unbraided, Arthur Clennam, trying to discover the whereabouts of Blandois (a.k.a. Rigaud), goes to Calais, where he has learned Miss Wade resides. He knows Miss Wade knows Blandois. Miss Wade is predictably perplexed. Why should Arthur Clennam 'press an undesired interest...in [her] affairs'? 'What can [he] have to do with the name [Blandois]?' (16.495). She knows nothing of Blandois's visiting Clennam's mother or his subsequent suspicious disappearance. Neither is she aware as yet that Arthur's pursuit pivots on a locked box that, it will later come out, Rigaud has left in her safekeeping—left with her precisely because she has nothing to do with it. All she knows of Rigaud is 'That he is a low, mercenary wretch; that [she] first saw him prowling about Italy...and hired him there' (16.497): 'A chance acquaintance, made abroad!' (16:498). Miss Wade will at last explain her involvement with Blandois–Rigaud, revealing her past affair with Henry Gowan and explicating the romance plot along with her life in 'The History of a Self-Tormentor'. (As it brutally turns out, the romantic pairings have required a full quad of breaking someone else's heart—Amy's of John Chivery's, Arthur's of Flora's, Pet's of Arthur's, Henry Gowan's of Miss

Wade's.) Arthur will have no reaction to Miss Wade's revelation. Why should he? This secret romance plot is not his plot. Or more correctly, as he would put it, actively suppressing his unrequited love for Pet, he is 'nobody' to it, and yet—this is the set-up—he knows more about it than any of its principal actors.

In the midst of the subsequent denouement of the locked-box plot, which *is* Arthur's direct concern, Dickens carefully marks the separation between the novel's two main plots. In an explicit pause and break in which Clennam and the rediscovered Rigaud await the response of Mrs Clennam to Rigaud's ultimatum setting her a one-week deadline to pay up, Rigaud tries to torment Clennam with separate knowledge he thinks Clennam does not know he has. 'I felicitate you on your admiration' for Pet (18.567), he mocks in characteristic fashion. ('Rigaud': from the French 'rigoler', to laugh, plus the English 'goad'.) But Clennam reveals in reply that he knows Rigaud has sold out Henry Gowan to spy for Miss Wade, and as Rigaud immediately deduces: 'I perceive you have acquaintance with another lady . . . Wade' (18.567). Rigaud then reconfirms the details of his hiring by her to Clennam, recapping the recipe of the romance plot, but 'he was saying nothing which Clennam did not already know' (18.568). Arthur Clennam is puzzling out another, separate set of interconnections.

What the protagonist Arthur Clennam does not know, and will still not know when the story ends, about this other, locked-box plot could almost fill a book about his not discovering it in which he plays a co-starring role—call it *Little Dorrit*. In brief, Arthur will not know literally the first thing about the authoring of his being by his biological mother, much less the sad story, which explains his grim upbringing, in which his father's clandestine love affair was discovered by the religious wife whom he later took in an arranged marriage—all, that is, that is evidenced by the terrible letters, secreted in the box, from his biological mother to his adopted mother begging for her baby back and for forgiveness. Nor will Arthur know that to keep this secret, Mrs Clennam has had to conceal a codicil to a will that reveals a tie between the Clennams and the Dorrits. This codicil, made by Arthur's father's uncle, the patriarch who helped keep Arthur's father and his biological mother apart, bequeathed a small but not insignificant amount of compensatory money to Arthur's biological mother and also to either the youngest daughter of her kind patron, who was Frederick Dorrit, or (in a formulation that feels a bit less strange to those familiar with lawyers' stratagems), if Frederick were childless and dead, to his brother's youngest daughter (Little Dorrit)—all contingencies that Dickens works out subsidiary—codicillary!—to establishing the second basic locked-box plot point: that the Clennams are '*connected* with the Dorrits' (memorandum).

The point here is that the vortex of this protagonist's ignorance thereby includes the most basic facts in the history of his family along with his family's buried connection to his future wife. Meanwhile, the knowledge that he, along with the reader, has about the romance plot never reaches Pet or her parents, whom it most deeply concerns. Little Dorrit, who met Rigaud in Italy on the romance plot, will never, as far as the reader knows, know that she also separately knew 'the man' who was Mrs Clennam's blackmailer (19&20.599). Dickens is careful never to name him to her, and, having left a packet at the prison containing copies of the contents of the locked box for her and Arthur, Rigaud is crushed to atoms by the collapsed house by the time she arrives on the scene. This novel makes it dangerous to use visual metaphors, but one might say that, when, at the end of the novel, Dickens throws some cross-lighting on the two main plots, the shadows only deepen for all the featured characters, while the murky penumbra of relevant information widens all around.

This final, differential semi-processing is fully to be expected. Dickens has all along been showing his characters journeying around a world in which the density and complexity of their interconnections, multiplying with every step, perpetually outruns its comprehension, such that they regularly encounter strangers who know things about themselves that they do not, while there is nothing like friends and family to highlight one's manifold obliviousness.

Indeed, encountering the potentially superior awareness of others is necessarily equally ubiquitous, and one achievement of this novel is to field each character as potentially setting off every other in this regard. In *Little Dorrit*, such higher-grade scaled knowing most baldly takes the shape of a slew of characters behaving as, or resembling, 'oracles' because in some relation to someone else they hold a comparatively omniscient-like perspective.

Little Dorrit offers a regular survey of pseudo-prophesying. One only has to look to start seeing it all over. Rigaud wields it: 'Do you ever have presentiments, Mr. Flintwinch?' he will tyrannize, withholding from Flintwinch the reason for his presence and his intentions; 'I, my son, have a presentiment to-night that we shall be well acquainted. Do you find it coming on?' (9.268). Pancks goes on for chapters playing at it. 'I am a fortune-teller. Pancks the gipsy' (7.212), he declares to Little Dorrit, pretending to read her palm, when in reality he is painstakingly piecing together the genealogy that will lead to her golden legacy. Little Dorrit herself is not above it. She subsequently replays Pancks's precognitive punning in reverse. 'Shall I tell you what my fortune is? And are you sure you will not share it?' Amy nudges Arthur, who does not yet know of his lover's relapse into poverty (Memorandum; 849). And while I am on the topic of fortunes, there is the global financier Merdle: 'here [Merdle] looked all over the palms of both his hands as if he were telling his own fortune' (17.530). He alone knows the calamitous nature of his own hidden complaint—that he is a swindler—and this ominous state of affairs becomes downright Damoclean when Merdle covertly alludes to his impending exit-by-suicide. 'Could you lend me a penknife?' he memorably requests; 'I'll undertake not to ink it' (17.530). Even the omniscient narrator practices the hocus-pocus. Recall the soliloquy of predictions the narrator conjures about a traveler coming to Mrs Clennam's—'Strange, if the little sick-room fire were in effect a beacon fire, summoning some one...' (5.129). As we saw, this prophesy actually registered that this traveler (Rigaud) is, unbeknownst to the reader, elsewhere discovering that which undergirds the prediction of his arrival (he is learning the contents of the locked box). Seemingly prescient omniscient narration turns out to be the exercise of the storyteller, carefully withholding information from the reader.[40]

Nor does Dickens always tip the hand wielding its disproportional awareness. In the character called only Mr F's Aunt, a sibylline discourse proceeds astoundingly in the face of total indeterminacy about what the seer knows or does not know. Without a doubt, Mr F's Aunt's every mad—Delphic—pronouncement, sagely interpreted, slams Arthur's reprehensible behavior toward her caretaker and his erstwhile lover Flora. (Predictably, my favorite is: 'There's mile-stones on the Dover road!', 7.196.) What, then, is the reader to make of this unnamed cryptic with the inscrutable face who rides in a glass coach and seems not to know what she is saying, when what she says has such well-targeted meanings? That question could be posed a dozen different ways, all unanswerable, and they would all apply equally well to the character Maggy, whose tragic early brain-damage and resulting arrest in perpetual childhood does not prevent her from displaying 20/20 second sight about the future that Little Dorrit portends for herself one time in the form of a fairytale. To the extent that one thinks of Maggy or Mr F's Aunt as intending their truths, they will likely line up with the other ersatz visionaries. To the extent that one accepts that they have a loose grip on reality, their divinizations will likely seem to descend from elsewhere—from the interpreting reader's capacious ascriptions of meaning; from the author who stands behind the characters' words; from the fact that their words, like all words, enact meanings syntaxed by linguistic and social forces that lie beyond any individual's consciousness.

It's a gallery of characters where one never knows who knows what. Beyond the pretend prophets, who hide what they know, and beyond the cryptic seers, about whom one never knows what they know, there stand, however, two others that are different. One is the servant Affery—an 'oracle', she is called (19&20.585). The other is Miss Wade. Both further complicate the semi-omniscient form of knowing through which the characters project views of themselves and others moving through the world.

First, Affery. In a series of similarly titled chapters, which are scattered across the novel and set off on their own, Affery witnesses a series of important plot scenes. She claims, however, to view these scenes, as the narrator does as well, as taking place in 'dreams'. Undercutting details and flatly contradictory narratorial remarks—'it was not at all like a dream' (1.30)—signal that they are no such thing. Within the story world, her husband Flintwinch helps justify Affery's denial of the reality she observes when, at the end of each such scene, he beats Affery into accepting that she must have been dreaming, and thus explains after the fact the narrator's focalization of her having 'dreamed' the scene. The reason why Flintwinch can believably do so, and a deeper reason that Affery and the narrator frame these scenes as 'dreams', is that neither Affery nor the reader can comprehend them. It is not that anything truly unbelievable happens in these scenes. On the contrary, notwithstanding that the nocturnal materialization of twins is pretty uncanny, Affery's dreams are only too realistic. Rather the reason these scenes can get re-classified as 'dreams' is that they do not fit into that recounting of happenings in time and space called the plot.

One will—even must—identify a dream, and even dismiss something that one took for reality as nothing but a dream, if whatever one believes one has observed—people, events, whatever—cannot be integrated into one's comprehension of time and space beyond the moment of its appearance. This is why one may coordinate having had a dream into reality, but it is nonsense to try to coordinate the events or the people occurring inside a dream with those occurring in reality. Rather what defines being inside a dream is that the dream people, who may perhaps be people one knows, and the dream events, familiar or not, cannot at another time be coordinated back into a time and space that one recognizes as going on diachronically and synchronically separately and outside oneself—and, therefore, define as real.[41]

In just this sense, Affery's false dreams reveal something about the plottable construction of reality. When, at the end of the novel, Affery patches her 'dreams' back into the plot, she re-coordinates the scenes she has witnessed into a plot that, while it concerns individuals, is made omnisciently real precisely in becoming untethered to her—or any individual, solely. These things happened not inside Affery, but outside her: her 'dreams' paradoxically retake their place as her own actual real subjective experiences only once they no longer belong to her personally, but rather can be integrated into an unindexed, coordinated time and space of a plot through which all the characters move.

Affery thus runs the pseudo-prophesying backward. Instead of her visions turning out to actually be indexed to her personally, something she knew but didn't admit to knowing, the narrative first falsely indexes her dreams to her as solely subjective and then makes her visions, conversely, omnisciently real. As Rigaud jeers, 'All that she dreams comes true' (19&20.591). In this instance at least, omniscient narration thus ought not to be described, as it is often rightly described, as arising intersubjectively from the aggregating, or consensus, of different individual perspectives. There must be a jump away altogether from individuals' spacetime to an unindexed coordinated time space.

No one can actually make that leap. All Affery or any individual can do is weave his or her experiences into it and recognize its theoretical, infinite existence. Where, then, one might wonder, does that leave Miss Wade's prediction? That is, how and why, in this theater of pseudo-prophets, does that other prophet, Miss Wade, come to make at the outset of the novel a real prediction that comes true? As she declares, of 'the men and women already on their road, who have their business to do with *you*' someone may be 'coming, for anything you know, or anything you can do to prevent it, from the vilest sweepings of this very town' (1.19). As it turns out, there is a vile prisoner in jail across the harbor—Rigaud—who does come and do business with both Pet and herself. How can Miss Wade be right? Or rather, what does it mean that Dickens structures the plot to make this prediction come true?

The answer lies, I believe, in what's changed historically, with what's now unprecedentedly true, about journeying. It is not merely that the numbers of people journeying on the roads has exceeded an amount where any sweeping gesture toward the people in any city is ample enough to take in someone that one will later encounter elsewhere. Even more importantly, the people journeying no longer represent individual travelers wending their various, sometimes crisscrossing, individual ways. They are moving as part of a *system* premised on their continuous circulation.

Miss Wade's prediction is the kind of prediction that one can make in the context of a system. Her prediction follows from the system's ongoing functioning and the resulting probabilities: she 'can be sure' that someone 'may' be coming. By contrast, imagine Miss Wade's prophesy transplanted into an ancient, Homeric epic. There the unwitting truth of her prediction would read as irony. The gods have their sport; the characters' limitations are unmasked. In *Little Dorrit*, it's completely different. Miss Wade is not ironized. The depth of Miss Wade's understanding is confirmed by the accuracy of her prediction, and what she is right about is what the journeying plot structure, which she crucially first invokes here, goes on to show: that the density of people's interconnections perplexingly exceeds their comprehension, that they never will catch up with what's happened sufficiently to understand what's happening, but that they can know that they are crisscrossing and moving in network. The international scope of a passenger transport system helps leverage this insight not just for Miss Wade, but for Dickens as well. It enables the author to abstract the system as a system existing beyond any individual's or any particular culture's possibility of comprehension.

The fact that Miss Wade makes her prediction within a fiction importantly divorces that prediction from the possibility of its being empirically proven or disproven. On the one hand, its fictional status likely inclines readers to say that it comes true because Dickens makes it come true. On the other hand, and at a deeper level, within a fiction Miss Wade's prediction gets to count as something more than an

assertion that the sort of thing that happened once may happen again—
a likely future spun from a known historical past. By not predicating
Miss Wade's prediction on any real past happenings or experiences, it
becomes—along with the story that enacts it—instead a reading of the
networked system's workings invoked by this 'fellow traveller'.

Miss Wade's omniscient-like journeying speech thus stands at the
outset of the novel for the relation of the individual to the system
producing this networked community of 'fellow travellers'. At the end
of the novel, in its final lines, Dickens expresses the obverse relation.
He foregrounds the omniscient narration's relation to the individuals
in the street. After Arthur and Amy's wedding ceremony, the narrator
concludes:

They [Amy and Arthur] paused for a moment on the steps of the [church's]
portico, looking at the fresh perspective of the street in the autumn morning
sun's bright rays, and then went down.
    Went down into a modest life of usefulness and happiness. Went down to
give a mother's care, in the fulness of time, to Fanny's neglected children no
less than their own. . . . Went down to give a tender nurse and friend to Tip for
some few years. . . . They went quietly down into the roaring streets, inseparable and blessed; and as they passed along in sunshine and in shade, the noisy
and the eager, and the arrogant and the froward and the vain, fretted, and
chafed, and made their usual uproar. (19&20.625)

Initially, here, Arthur and Amy descend into a literal street. Then, with
the repetition of the end-phrase 'went down' at the beginning of the
next line, their movement gets re-signified as figurative of their future
lives, with 'went down' opening three sentences that broadly encompass things not literally exclusively done in the street: working, raising
children, looking after a troubled sibling. In the final sentence, this
figurative procession down the street then gives way again to the literal
streets but, at the same time, saturates those streets with its totalizing
figurative meaning. The 'as' of 'as they passed along' puddles their
movement into unspecified time that 'in the sunshine and in the shade'
further diffuses, while the nominalized adjectives—e.g. 'the noisy and
the eager'—generalize the human crowd through which they move,
re-enforcing the smudging effect of pluralizing 'the street' into 'the
streets'.

In a sense, one job of Dickens's omniscient narrator has been to sup-
ply this perspective that pincers the characters between an account of
their literal movement down streets and its totalizing symbolic

figuration, to which that emplotted movement gives rise, of their journeying. Unlike in *Master Humphrey's Clock*, these characters as they journey self-awarely imagine their plottability in relation to an unattainably comprehensive perspective. They know they will be thinking and re-thinking of the going as long as they go. In this sense, readers really have been reading a story about 'looking at the fresh perspective of the street', nor should they resolve the ambiguity and odd redoubling of vision around whether that perspective takes in the street or is being ascribed to it.

Notice as well, though, that the omniscient perspective takes shape around the act of selecting a pair of fellow-travelers whose paths readers have seen intersect and who are now continuing on their journeying together through the crowded frenzy—'the roaring streets' in an 'uproar'. Such an omniscient perspective might perhaps seem at first simply to recount the pair's movement in tandem with others through space over time, observing Arthur and Amy '*as they passed along in sunshine and in shade*'. In that view, the sun's movement, and the time it keeps, is thought to produce the alternating light and shadow on the two as they pass along. But that is not exactly how Dickens pictures it. Rather, as the narrator observes, the pair's movement as they pass along the street creates an alternation of sunshine and shade on them. Their movement, not the sun's, makes time here.

Sunshine and shade: Dickens is circling his readers back to the beginning of his story, back to his first chapter, 'Sun and Shadow'. Recalling that strange opening onto the deformation of perspective, and the way in which it wrested the formation of the omniscient narratorial perspective from the sun and the time it keeps, he recalls also how in the sun's place the omniscient narrator's perspective formed around the crisscrossing relations of 'fellow travellers' on their journeys. This involved a basic shift in the formation of perspective: recentering it on the networked mobility of people, which was helping reconvention time itself in Dickens's time.

It's a 'quiet conclusion' (memorandum). Yet it is perhaps no less powerful for that. Treating it also as a means to concluding this study about how the passenger transport revolution went down in sunshine and shade, one perhaps gets to discover that Charles Dickens was that revolution's Copernicus. In *Little Dorrit*, he was no longer elatedly presenting the convergence of the networked community as he had pictured it in Pickwick's stage-coaching adventures. Nor was he rendering its tragic breakdown, its limits, as he had done in *Master Humphrey's Clock*. Those were key moments of evolving understanding. In this autumnal moment, however, Dickens seems to have comprehended that what his art—the art of the novel—could show about the passenger transportation revolution and its networking of people was, more mutedly, the surprising changes in perspective that it was bringing.

Afterword

Dickens writes so cogently about passenger transportation systems that it has been something of a struggle not to extend this book's arguments distractingly to his other novels and their many intense scenes of mobility. In *Nicholas Nickleby* (1838–9), for instance, where Dickens repeatedly separates characters from each other to test the logic of their isolation and connectedness, the Snow Hill coach and London's Saracen's Head Inn stand at the story's outset for both, with Squeers vanishing young Nicholas away to Yorkshire. (Dickens crafts, I believe, a multiplotted road novel to resolve this story's impasse between comedy and tragedy.) Omitted also in despite of its obvious relevance is *Martin Chuzzlewit* (1843–4), in which Dickens sends Selfishness on a transatlantic round trip. Most of all, however, it has seemed a risk not to offer up an analysis of *Dombey and Son* (1846–8). Carker's flight, in particular, stunningly links the stage coach to the railway, and, as in *Dombey* as a whole, raises the problem of one's determination by, and escape from transport systems, with their collectivizing and time-warping force. The passenger transport revolution is also—in my view—the revolution unfurling behind the French republican one depicted in *A Tale of Two Cities* (1859). Where the first chapter announces that 'It was the best of times, it was the worst of times . . .' and ponders how to define 'The Period' (the chapter title), the second chapter answers: 'It was the Dover road . . .', invoking the road connecting London out to the Continent over which history had come until the railways took another route and made that road itself historical. On that rail route, in 1865, Dickens's train crashed at Staplehurst, and Dickensians will have missed a discussion of this accident and of the postscript to *Our Mutual Friend* (1864–5), where the characters in manuscript are described as surviving it. One might easily go on noting other pertinent work by Dickens. There is, however, perhaps a more important question.

Why only Dickens? What generalizability, what applicability do the arguments have that this book has made about just three of Dickens's novels? How are, for instance, novels by Scott, Brontë, Gaskell, Thackeray, Trollope, Collins, Hardy, and so on, similar or different? What about Twain, Zola, or Gogol? Moreover, in mounting arguments about Dickens's achievement as a novelist and his place in the history of his culture and possibly our own, this book frequently signals that his fiction was giving his audience, and us, a set of ways of understanding the world in the age of passenger transport networks. What wider conclusions ought, then, a reader to draw?

One chronological continuation of this book might have discussed Jules Verne's *Around the World in Eighty Days* (French serial 1872, English 1874) as a novel that pays homage to *Pickwick*, then Joseph Conrad's *Secret Agent* (1906–7) for the bombing of Greenwich time, and finally E. M. Forster's *Howard's End* (1910) where the automobile confronts the imperative to 'Only connect...'. And, given world enough and time, why stop there? Disembarkation from an airplane, for instance, serves memorably as the denouement to James Baldwin's *In Another Country* (1962), while a mid-air explosion and plunge dramatically open Salman Rushdie's *The Satanic Verses* (1988). The afterword to such a book could have taken in the GPS technology currently transforming everyday life. At the same time, broadening away from Dickens equally brings into view not only that which follows him, but also the selectivity that kept out contemporary works. R. D. Blackmore's *Lorna Doone* (1869) probably takes the prize for a nineteenth-century novel that looks back from the expanded range and general mobility enabled by a public transport revolution to an earlier localism and emplacement, sleds in a valley. Not so differently, the imagined historical isolation of *Wuthering Heights* (1847), beginning with Earnshaw's sixty-mile walk to Liverpool in 1771 where he finds Heathcliff, garners some of its meaning from the fact that railways were carrying this novel's readers. Even the briefer twenty-year backward glance of *Jane Eyre* (1847) deserves re-reading from the perspective of transport history, and it has notably received it in a fine recent article (which ties it to national politics).[1] Why, it might also have been asked in a more expansive version of this book, does the mystery of Wilkie Collins's *The Woman in White* (1860–1) pivot on discovering when Laura caught the train to London and took a cab, whose driver remembers her? Does the Grand Trunk Road in Rudyard Kipling's *Kim* (1901) cut the Gordian knot tied by the Gate of the Way and the Great Game of Empire? Moreover, what, one might wonder, about other national literatures? In *Le Rouge et le Noir* (1830), Stendhal declared that 'a novel is a mirror on a highway', while for Mark Twain that highway was the Mississippi. Gogol memorably compared Russia to a troika in *Dead Souls* (1842): 'And you, Rus, are you not also like a brisk, unbeatable troika racing on? The road smokes under you, bridges rumble, everything falls back and is left behind.'[2] Famously, Émile Zola's *La Bête Humaine* (1890) makes its subject and structure the train network as a metonym for French society generally.[3] In a sense, in conjuring these other novels, we come full circle back to the moment in the Introduction when George Eliot's *Felix Holt* and Thomas De Quincey's *The English Mail Coach* were mentioned only to be dismissed for a focus on Dickens.

At the same time, however, this surveying glance might also call to mind once again the 'March of Intellect' engraving that serves as this book's opening illustration. There too capturing public transport's busyness meant risking missing its deeper coherences. The problem is not merely that a surveying approach is inevitably provocative as much for that which it leaves out as for that which it includes. An inherent epistemological challenge lurks in how to write, on the one hand, about a closed system in an unfettered way and, on the other, about an expansive and expanding network in a focused analytical mode, rather than in a panoramic and associative one. In trying to meet that challenge, initially it simply helped to focus on the evolving writings of a single rider. Gradually, though, a logic clinching this focus on an individual passenger also emerged. It was that individual passengers alone offered the possibility of comprehending both the network and themselves through learning to imagine their journeys from a third-person narrative perspective.

But at my back I always hear
Time's winged chariot hurrying near;
. . .
Now, therefore. . .
Let us roll all our strength and all
Our sweetness up into one ball,
. . .
[For] though we cannot make our sun
Stand still, yet we will make him run.

Andrew Marvell

But at my back from time to time I hear
The sound of horns and motors. . .

T. S. Eliot

Notes

INTRODUCTION

1. George Eliot, *Felix Holt, The Radical* (Edinburgh: Blackwood, 1866), quotations are from pages 9, 3, and 7, but see Eliot's whole 'Introduction'.

2. Charles Dickens, *Little Dorrit*, 20 numbers in 19 monthly parts (London: Bradbury & Evans, 1855–7), 1.19. All citations to *Dorrit* are to the part number and page.

3. Throughout this book I draw continually upon the rich and fascinating work of historians of British transport, and at the outset I want to acknowledge several cornerstone studies, beginning with two towering early tomes: W. T. Jackman's epochal *The Development of Transport in Modern England* (Cambridge, 1916; rpt. New York, 1965) and Edwin A. Pratt's *A History of Inland Transport and Communication* (London, 1912; rpt. New York, 1970). In addition to Philip S. Bagwell's *The Transport Revolution* (London: Routledge, 1974), a general transport history book I found especially useful in a day-to-day way is P. J. G. Ransom's *The Archeology of the Transport Revolution, 1750–1850* (Tadworth: World's Work Ltd., 1984). Ransom nicely charts the simultaneously interweaving strands of different forms of transport. In subsequent notes, I offer brief, selective bibliographies of passenger transport history and theory subdivided by relevant modes and means of transport, i.e. stage coaches, roads, horses, railways, ballooning, cabs, canals, pedestrianism, steam ships. These mini-bibliographies can also be located using the index.

4. See Stephen Kern's *The Culture of Time and Space, 1880–1918* (Cambridge, Mass.: Harvard University Press, 1983) for an approach that enlighteningly organizes conceptually around time and space. Kern also treats the history of standardized time, dating it later than I do as his subtitle indicates. He argues, mistakenly I believe, that modernist art pitted psychological time against the 'linear' form of standardized time.

CHAPTER I

The Speeding of the Pickwick Coach

1. Charles Dickens, *The Posthumous Papers of the Pickwick Club*, 20 numbers in 19 monthly parts (London: Chapman & Hall, 1836–7), 1.1. All citations

to *Pickwick* are to the part number and page. '1827' was initially misprinted as '1817', and Dickens actively corrected this misdating; the corrected date—taken from the serial's errata—appears in all subsequent editions and has been silently incorporated here.

2. Steven Marcus, *Dickens from Pickwick to Dombey* (New York: Simon and Schuster, 1965), 17. The novel's comic aspect has been explained as delivering a 'transcendent' and 'supernatural' modern Eden of secularized Christianity (Steven Marcus, W. H. Auden); as releasing regenerative forces through transmuting the energy of ostensibly low, ephemeral popular culture into legitimate art (Paul Schlicke); as shining the heroics of a latter-day Don Quixote upon a knightless age (Alexander Welsh); and as testifying to charmed benevolence surviving in, or adapting to, a modern inhumane world (J. Hillis Miller, James Kincaid).

3. An exemplifying use of the cultural catchphrase 'the annihilation of space and time' can be found in *Fraser's* 'Railways' (July 1838), 47–8; in 1833, *The Railway Companion, describing an excursion along the Liverpool line* (London) was already describing how the effect was 'to annihilate—or, at least immeasurably extend—the bounds of time and space (p. 16).

4. James Kinsley, 'Introduction' in Charles Dickens, *The Pickwick Papers*, ed. (Oxford: Clarendon Press, 1986), p. lv. The view of *Pickwick*'s coaching as nostalgic corrupts even Humphry House's stellar chapter on history in *The Dickens World* (Oxford: Oxford University Press, 1941); House otherwise provides the authoritative starting point for considerations of Dickens and stage coaching. For a recent discussion of *Pickwick* in relation to coaching, see Gina M. Dorré, *Victorian Fiction and the Cult of the Horse* (Aldershot: Ashgate, 2006), ch. 1: 'Handling the "Iron Horse": Dickens, Travel, and the Derailing of Victorian Masculinity'.

5. John Forster, *The Life of Charles Dickens*, ed. J. W. T. Ley (Cecil Palmer, 1928), 61.

6. Dickens to Thomas Beard, 2 May 1835, in *The Letters of Charles Dickens*, ed. Madeline House, Graham Storey, and Kathleen Tillotson, 12 vols. (Oxford: Clarendon Press, 1965–2002), 1.58.

7. Peter Lecount, *A Practical Treatise on Railways, Explaining Their Construction and Management* (Edinburgh, 1839), 196. The problem of communication with the driver was partly resolved by introduction of the emergency brake in 1889; see Jack Simmons's *The Railway in England and Wales, 1830–1914* (Leicester: Leicester, 1978), 223–31.

8. Paul Virilio, *The Art of the Motor*, trans. Julie Rose (Minneapolis: University of Minnesota Press, 1995), 49.

9. Kathryn Chittick, *Dickens and the 1830s* (Cambridge: Cambridge University Press, 1990), 61–72; John Butt and Kathleen Tillotson, *Dickens at Work* (London: Methuen, 1957), 73–4. The story opens in May 1827 and closes October 1828 (with the November number).

10. See Malcolm Andrews, 'Dickens, Washington Irving, and English National Identity', *Dickens Studies Annual* 29, 1–16.

11. Wolfgang Schivelbusch, *The Railway Journey: The Industrialization of Time and Space in the 19th Century* (Berkeley: University of California Press, 1986), 17–18.

12. On the history of coaching, see W. T. Jackman, *The Development of Transport in Modern England* (Cambridge, 1916; rpt. New York, 1965); Edwin A. Pratt, *A History of Inland Transport and Communication* (London, 1912; rpt. New York, 1970); Philip S. Bagwell, *The Transport Revolution* (London: Routledge, 1974); P. J. G. Ransom, *The Archeology of the Transport Revolution, 1750–1850* (Tadworth: World's Work Ltd., 1984). G. A. Thrupp's *The History of Coaches* (Amsterdam: Meridian, 1969, first published in 1877) offers a fine history written in the nineteenth century. There are many useful lay histories of coaching, notably, R. C. and J. M. Anderson, *Quicksilver: A Hundred Years of Coaching, 1750–1850* (Newton Abbot: David & Charles, 1973) and David Mountfield, *The Coaching Age* (London: Robert Hale, 1976). *Carriage Terminology: An Historical Dictionary* (n.p.: Smithsonian Institution, 1978), compiled by Don H. Berkebile, offers exactly the resource its title promises, and it is filled with helpful illustrations of the many different vehicles. The earliest days of stage coaching, that is, the bare beginnings of the network before even the eighteenth-century turnpikes, are recovered by Dorian Gerhold in *Carriers and Coachmasters: Trade and Travel before the Turnpikes* (Chichester: Phillimore, 2005).

13. *Sketches by 'Boz', Illustrative of Every-day Life, and Every-day People,* 2 vols. (London: Macrone, 1836), 2.171. Citations to the 1836 edition are to volume and page.

14. On the evolution of road technology, see Graham West's *The Technical Development of Roads in Britain* (Aldershot: Ashgate, 2000). Arthur Young measured the four-foot-deep mud ruts in his *Tour in the North of England* (vol. 4 [1770], 580; for a view contemporary with Dickens's, see G. R. Porter, *The Progress of the Nation* (London, 1838), ch. 2: 'Turnpike Roads'.

15. On the turnpike system, see especially Jackman, *Development of Transport*, ch. 4, 'Roads and Their Improvement, 1750–1830'; William Albert, *The Turnpike Road System in England, 1663–1840* (Cambridge: Cambridge University Press, 1972); John Copeland, *Roads and Their Traffic, 1750–1850* (Newton Abbot: David & Charles, 1968), and W. J. Reader, *Macadam: The McAdam Family and the Turnpike Roads, 1798–1861* (London: Heinemann, 1980).

16. *The Horse; with a Treatise on Draught,* Library of Useful Knowledge (London: Baldwin and Cradock, 1831), 35. Juliet Clutton-Brock recounts *A History of the Horse and the Donkey in Human Societies,* as advertised by her subtitle to *Horse Power* (Cambridge, Mass.: Harvard University Press, 1992). See also Richard W. Bulliet, *The Camel and the Wheel* (Cambridge, Mass.: Harvard University Press, 1975).

17. Bagwell, *Transport Revolution,* 29. Such numbers are always in dispute; for instance, see Theo Barker and Dorian Gerhold on speeds in the eighteenth century in *The Rise and Rise of Road Transport, 1700–1990* (Basingstoke:

Macmillan, 1993). (They find greater acceleration in the 1750s and 1760s than in the 1800s.) Bagwell's figures accord well with what seems to be Dickens's perception.

18. One eighteenth-century equivalent of the Rainhill trials was the bet won on 29 August 1750 by the Duke of Queensberry: he apparently covered 19 miles in a four-wheeled machine in fifty-four minutes, a rate of just under 19 m.p.h. See William Kitchiner, *The Traveller's Oracle; or, Maxims for Locomotion*, 2 parts (London: Colburn, 1827), 2.221; *Gentleman's Magazine* (1750), 379, 440.

19. Bagwell, *Transport Revolution*, 37.

20. Quoted from *Huntingdon, Bedford, Cambridge, & Peterborough Gazette* 15 May 1830, in Copeland, *Roads and Their Traffic*, 101.

21. Four excellent histories of the railway system that attend to its cultural and public aspects are: Jack Simmons's *The Victorian Railway* (New York: Thames and Hudson, 1991); Schivelbusch, *The Railway Journey*; David Turnock, *Railways in the British Isles: Landscape, Land Use, and Society* (London: Black, 1982); and Michael Freeman, *Railways and the Victorian Imagination* (New Haven: Yale University Press, 1999). On the railways' transformation of time and space see especially ch.2 in Freeman, *Railways*, titled 'The "March of Intellect" '. Simmons, Schivelbusch, and Freeman all also discuss reading on the railways and the establishment of station book stalls by W. H. Smith; the renaissance of book history has garnered this topic much attention.

22. 'The Manchester and Liverpool Rail-road', *The Penny Magazine of the Society for the Diffusion of Useful Knowledge* 69 (1833), 167.

23. Humphry House first remarked that Dombey rides in his private coach strapped onto an undercarriage (in *The Dickens World* (Oxford: Oxford University Press, 1941), 140), and John Sutherland further explores and explains in 'Visualizing Dickens', in John Bowen and Robert Patten (eds.), *Palgrave Advances in Dickens Studies* (Basingstoke: Palgrave, 2006).

24. F. M. L. Thompson, *Victorian England: The Horse-Drawn Society* (London: Bedford College, 1971). Thompson first called attention to a widespread twentieth-century misconception that railways displaced horses in Victorian England. In this regard, he rightly chastised 'the nostalgic...school of historians [who had] taken its cue from Dickens and *Pickwick Papers*' (p. 13), but he should have added as well that they were misreading *Pickwick* as nostalgic. John Poole's 1835 tale titled 'The Inconveniences of a Convenient Distance' in *Sketches and Recollections* (London: Colburn, 1835) nicely illustrates the intensifying networking by horse.

25. B. W. Matz, *The Inns and Taverns of 'Pickwick'* (New York: Scribner, 1922), 178; according to the *Directory of Stage Coach Services 1836* compiled by Alan Bates (Newton Abbot: David & Charles, 1969), by 1836 the coach to the White Hart at Bath is no longer run by Pickwick. Joseph Gold, in

Radical Moralist (Minneapolis: University of Minnesota Press, 1972), appears to be the first to connect the full name 'Moses Pickwick' to the fact that the name 'Boz' came from a comic nasal pronunciation of 'Moses'.

26. House, *The Dickens World*, 21–8.

27. Reinhart Koselleck, *Futures Past: On the Semantics of Historical Time*, trans. Keith Tribe (New York: Columbia University Press, 2004); the 'contemporaneity of the noncontemporaneous' is a repeated phrase, but see especially p. 246. Also pertinent is James Chandler's historicization of early nineteenth-century historiography as the 'Age of the Spirit of the Age' in *England in 1819: The Politics of Literary Culture and the Case of Romantic Historicism* (Chicago: University of Chicago Press, 1998).

28. See Georg Lukács, *The Historical Novel*, trans. Hannah and Stanley Mitchell (Lincoln: University of Nebraska Press, 1962). Looking to France, Lukács sees Balzac as historicizing the present, and the July revolution of 1830 as a reason for its historicization.

29. Angus Fletcher offers a discussion of how motion itself first had to be historically imagined as a property immanent in nature in *Time, Space, and Motion in the Age of Shakespeare* (Cambridge, Mass.: Harvard University Press, 2007). On theorizing human motion, see Brian Massumi's *Parables for the Virtual: Movement, Affect, Sensation* (Durham, NC: Duke University Press, 2002).

30. Franco Moretti especially has drawn maps to open up both novels and genres in relation to history, and here I draw inspiration from his work. See 'Maps', in *Graphs, Maps, Trees: Abstract Models for a Literary History* (London: Verso, 2005) and also *Atlas of the European Novel, 1800–1900* (London: Verso, 1998), which not only discusses Dickens, but also provocatively charts the picaresque.

31. Specially relevant to *Pickwick* is Sydney Hall's map 'Inland Communication' published in John Gorton's *Topographical Dictionary* (Chapman and Hall, 24 monthly shilling parts, 1831–3); the narrator alludes to Gorton's book at the opening of ch. 13. Hall's four-plate, intensely detailed map shows proposed and existing railways, canals, and roads.

Network maps and timetables have an important design history not fully addressed in this book—or anywhere else that I have found. In brief, the eighteenth century saw the development of road books for travelers on the go—for wayfinding en route instead of advance planning. John Cary's road books especially marked an important shift in the 1790s by adding inns and turnpikes to the maps. *Paterson's Roads* was the *Bradshaw's* of the preceding age of road travel. Edward Mogg took up Paterson's mantle for a fast-coaching era (e.g. *Mogg's Paterson's Roads*), but he failed to recognize the new timetabling and network mapping requirements of the railways. Hence *Mogg's Handbook for Railway Travellers* (1840) fell into oblivion. *Bradshaw's Monthly Railway and Steam Navigation Guide* became the passenger's bible. Dickens comically dissects *Bradshaw's* in 'A Narrative of Extraordinary

Suffering' in *Household Words* (12 July 1851). Recently, Mike Esbester has broached the topic of 'DesigningTime:The Design and Use of Nineteenth-Century Transport Timetables' in *Journal of Design History* 22 (2009), 91–113. A beginning effort was also made with a collection of essays, edited by James R. Akerman, titled *Cartographies of Travel and Navigation* (Chicago: University of Chicago Press, 2006); Akerman's excellent introduction confirms and nicely clarifies how shockingly little work has been done on this important genre.

32. Hobbes's phrase, 'life it selfe is but motion', comes from his *Leviathan* (1651). When I use it, I intend both to gesture toward the original treatise's explicit defense of the necessity for citizens to subject themselves to a civil authority, i.e. a monarch, in order to live in community, and to pressure Mark Seltzer's modern citation of the phrase to hold together the relays between communication and transport systems (e.g. 'motion pictures').

33. See, for a very different but also very closely related historical viewpoint, Jeffrey Schnapp, 'Crash (Speed as Engine of Individuation)', *Modernism/Modernity* 6 (1999), 1–49. I am contending by contrast to Schnapp that in the public transit system speed becomes an engine of collectivity.

34. 'A General System of Railways for Ireland', *Quarterly Review* 63 (1839), 22.

35. G. K. Chesterton's quotation, long a staple of bookjacket blurbs, comes from his influential and impressive *Charles Dickens: The Last of the Great Men* (New York: Readers Club, 1942 [first published 1906]), 59. Chesterton's title is not as old-fashioned as it perhaps sounds: the last great man is the one who sees, democratically, greatness dispersed everywhere. Standing at the other end of the century's Dickens criticism, John Bowen's excellent, Derridean-influenced analysis of *Pickwick* is essential reading along with the rest of his *Other Dickens: Pickwick to Chuzzlewit* (Oxford: Oxford University Press, 2000).

36. How the imagined community constructed by the public transport system relates to Benedict Anderson's thinking about the imagined community of the nation is taken up directly in Chapter 3, 'International Connections'. Readers familiar with Anderson's work may already have begun, however, to discern both the connections and challenges to Anderson. My use of the provocatively problematic term 'community' throughout this book is partly intended to keep Anderson's 'imagined communities' in view; see Benedict Anderson, *Imagined Communities: Reflections on the Origin and Spread of Nationalism* (London: Verso, 1983).

37. David Parker, 'Mr Pickwick and the Horses', *The Dickensian* 85 (Summer 1989), 88.

38. Hans Robert Jauss, 'The Comic of Innocence—Innocence of the Comic (Dickens's Comic Hero)', in *Aesthetic Experience and Literary Hermeneutics*, trans. Michael Shaw (Minneapolis: University of Minnesota Press, 1982), 213–20.

39. The Fleet episode draws on F.W.N. Bayley's *Scenes and Stories by a Clergyman in Debt* (London: Baily, 1835); James Kinsley sketches the overlaps in the Clarendon *Pickwick*, pp. lxxii–lxiii.

40. For an excellent elucidation of the largely unacknowledged difficulty that literary criticism has in breaking through to the real roles played by the novel, see Robert Newsom's *A Likely Story: Probability and Play in Fiction* (New Brunswick: Rutgers University Press, 1988).

41. [Abraham Hayward?], unsigned review, *Quarterly Review* (October 1837), 484.

42. Charles Dickens, 'A Flight', *Household Words* (30 August 1851), 533.

43. Dickens's obituary in *The Illustrated London News* (18 June 1870) offers a pithy summary of the widespread awareness that in all his novels 'the course of his narrative seemed to run on, somehow, almost simultaneously with the real progress of events': one 'obvious effect was to inspire all his constant readers . . . with a sense of habitual dependence on their contemporary, the man Charles Dickens' (p. 639).

44. Mark W. Turner, ' "Telling of my weekly doings": The Material Culture of the Victorian Novel', in Francis O'Gorman (ed.), *A Concise Companion to the Victorian Novel* (Oxford: Blackwell, 2005), 119. Turner briefly links serialization to standardized time as well. On serialization, Richard Altick's *The English Common Reader: A Social History of the Mass Reading Public 1800–1900* (Chicago: University of Chicago Press, 1957) is a standard reference book that lucidly lays out the social and economic conditions taken to govern serial and book circulation. In 1957, the importance of studying publishing and writing practices, especially serialization, also came home to Dickens studies with John Butt and Kathleen Tillotson's breakthrough *Dickens at Work*. These investigations laid the groundwork for even sharper studies of serial publication by a next generation of scholars, led by Robert Patten and John Sutherland. See especially Patten's *Charles Dickens and His Publishers* (Oxford: Clarendon Press, 1978) and Sutherland's *Victorian Novelists and Publishers* (Chicago: University of Chicago Press, 1976). Not surprisingly, out of such careful attention to authorial labor, publishing contracts, and circulation figures, the most provocative and important extended reading of the relationship between serialization and Pickwick that emerged was Marxist: N. N. Feltes's *Modes of Production of Victorian Novels* (Chicago: University of Chicago Press, 1986). A deconstructive and Benjaminian twist has been added by Kevin McLaughlin in *Writing in Parts: Imitation and Exchange in Nineteenth-Century Literature* (Stanford: Stanford University Press, 1995).

45. Turner, ' "Telling of my weekly doings" ', 116.

46. B. C. Saywood, 'Dr Syntax: A Pickwickian Prototype?', *The Dickensian* 66 (1970), 24–9.

47. William Combe, *The Tour of Doctor Syntax, In Search of the Picturesque* (London: R. Ackermann, 1812), pp. i–ii. Combe also used the just-in-time

publishing method for monthly parts to bring out his *The English Dance of Death* in 24 numbers (1814–16).

48. Both Robert Seymour and William Heath, the artist of the 'March of Intellect' illustration discussed in the Introduction, worked with the publisher Thomas McClean.

49. On the steam coaches, see Copeland, *Roads and Their Traffic*, ch. 7.

50. J. Hillis Miller, 'The Fiction of Realism: *Sketches by Boz, Oliver Twist*, and Cruikshank's Illustrations', in Ada B. Nisbet (ed.), *Charles Dickens and George Cruikshank* (Los Angeles: Clark Library, 1971), 1–69. Miller quotes as his epigraph Dickens's precisely bi-directionally ambiguous phrase: 'the illusion was reality itself'. Amanpal Garcha calls thoughtful attention to the formal import of transport to *Sketches* in *From Sketch to Novel: The Development of Victorian Fiction* (Cambridge: Cambridge University Press, 2009).

51. *The Times* (18 September 1834), 3.

52. On ballooning, see L. T. C. Rolt, *The Aeronauts: A History of Ballooning, 1783–1903* (Brunswick: Sutton, 1985). One can read all about *The History of Aeronautics in Great Britain from the Earliest Times to the Latter Half of the Nineteenth Century* in a wonderful, magisterial old book of that title by John Edmund Hodgson (London: Oxford University Press, 1924).

53. Dickens similarly positions 'you' on the stage coach in an important passage, not analyzed here, in ch. 16 of *Pickwick*; he also commonly uses journeying to shift into a present tense series, e.g. 'And now . . .'.

54. Later I address the distinction, collapsed here, between the virtual 'you' made to seem to travel by the communication system and the circulation of bodies in time and space by a passenger transport system. Alison Byerly takes up this subject in earnest from its communications side in *Virtual Travel and Victorian Realism* (forthcoming). See also Dickens's Uncommercial Traveller essay later titled 'Nurse's Stories', in which Dickens riffs on revisiting 'the places to which I have never been'—the imagined places in books (*All the Year Round* (8 September 1860), 517).

55. Reprinted in the Clarendon *Pickwick*, ed. Kinsley, 884–5.

56. *The Athenaeum* (26 March 1836), 232. The advertisement for *Pickwick* was published elsewhere as well; for instance, on 31 March 1836, concurrent with the first number of *Pickwick*, on the back wrapper of the first monthly number of *The Library of Fiction, or, Family Story-teller* (London: Chapman & Hall, 1836), which included Dickens's short tale 'The Tuggs's at Ramsgate', later republished in *Sketches*.

57. Turner, ' "Telling of my weekly doings" ', 116.

58. Thomas Keymer, *Sterne, The Moderns, and the Novel* (Oxford: Oxford University Press, 2002); see also Keymer's 'Reading Time in Serial Fiction before Dickens', *Yearbook of English Studies* 30 (2000), 43–5.

59. Letter from Mary Russell Mitford to Miss Jephson, 30 June 1837, in *The Life of Mary Russell Mitford, Told by Herself in Letters to her Friends*, ed.

A. G. K. L'Estrange, 2 vols. (New York: Harper, 1870), 2.198. See also George H. Ford, *Dickens and his Readers: Aspects of Novel-Criticism since 1836* (Princeton: Princeton University Press for the University of Cincinnati, 1955), ch. 1: 'The Prospering of Pickwick'.

60. See Sutherland, *Victorian Novelists*, and Patten, *Charles Dickens and His Publishers*. Catherine George Ward and Hannah Maria Jones published many a novel in weekly and monthly parts in the preceding decades, according to expert book historian Peter Garside in his and Rainer Schöwerling's *The English Novel 1770–1829: A Bibliographical Survey of Prose Fiction Published in the British Isles* (Oxford: Oxford University Press, 2000), 95–6.

61. Patten, *Charles Dickens and His Publishers*, 46.

62. Anny Sadrin, 'Fragmentation in *The Pickwick Papers*', *Dickens Studies Annual* 22 (1993), 24–5.

63. Miller, 'Fiction of Realism', 13.

64. Niklas Luhmann's *Social Systems* (trans. John Bednarz with Dirk Baecker, Stanford: Stanford University Press, 1995) has broadly influenced my understanding of passenger transport systems, and it here forms a theoretical backdrop. Systems theorists, though, are likely to be disappointed. I am not concerned with Luhmann's highly influential work on autopoesis, in which he rightly begins analyzing with the assumption that systems are self-observing and self-modifying. Nor, in that vein, do I treat, as one could, the passenger transport system as a system serving only its own evolving. Nonetheless, I have relied heavily on Luhmann. In particular, I too insist that systems be defined by identifying their functions and that they only come into view when one makes a distinction allowing one to explore that function.

Bruno Latour's work has chiefly shaped my thinking about networks— generically in *We Have Never Been Modern*, trans. Catherine Porter (Cambridge, Mass.: Harvard University Press 1993) but also more specifically in relation to public transit in *Aramis, or the Love of Technology*, trans. Catherine Porter (Cambridge, Mass.: Harvard University Press, 1996). From Latour, I learned not to make arguments that banally point to our technological hybridity, which has always been with us, and instead to enjoy the problem of how we see the collectivity of networks in which we are always local. As was mentioned in the Introduction, there is not a sharp differentiation made here between 'networks' (which term, especially in Latour, tends to emphasize connectivity and its extensions) and 'systems' (a term that tends toward enclosing, if shifting, boundaries and self-defining processes). Again, both aspects are seen as jointly in play. The fact that, unlike the post, the passenger transport system arose from many different private companies (such as Pickwick's), which could even aim at adversarial disconnectivity (most dramatically with competing railway gauges) only makes even clearer the form in which to distinguish the system as a system (e.g. not as an economic entity).

65. Howard Robinson's *The British Post Office: A History* (Princeton: Princeton University Press, 1948) is *the* history of the post. Bernhard Siegert's *Relays: Literature as an Epoch of the Postal System*, trans. Kevin Repp (Stanford: Stanford University Press, 1999) provocatively recounts the history of communications from 1784, tracing the way modern literature changes as communications increasingly come to be seen as a self-generating system with its meanings inhering in its shifting technological forms.

66. On epistolary form, see Janet Gurkin Altman, *Epistolarity: Approaches to a Form* (Columbus: Ohio University Press, 1982). My reading aims to address partly the need Altman sees 'for careful assessment of the changes in society and esthetic ideals that underlie the decline of a form such as the letter novel...[and to] bridge the gap between eighteenth- and nineteenth-century novel studies by investigating more thoroughly the evolutionary or dialectical relations obtaining between the two' (p. 195). Especially I am concerned with 'the phenomenon of 'presentification'...whereby the writer tries to create the illusion that both he and his addressee are immediately present to each other and to the action...[which] tendencies suggest an eighteenth-century reading public whose dominant esthetic is contemporaneity' (p. 202). See also Irene Tucker's analysis of epistolarity (which also quotes this passage from Altman) in 'Writing Home: *Evelina*, the Epistolary Novel and the Paradox of Property', *ELH* 60 (1993), 438.

67. I only wish I could claim credit for the wonderful discovery that 'Bardell' was the name of a transport firm, but the prize for this piece of sleuthing goes to William Long; see 'Mr Pickwick Lucky to Find a Cab?', *The Dickensian* 87 (1991), 167–70. According to Long, the Bardell line had switched over from hackney coaches to omnibuses and cabs by the 1830s.

68. Bowen, *Other Dickens*, 55.

69. As Mark Seltzer observes, when Pierre-Simon Laplace recognized the need to explain that the number of dead letters in the post office remained constant each year—since no one intends to write a dead letter—Laplace helped broach the conceptualization of systems 'irreducible to individual intentions'. On theorizing systems, see Seltzer's *Bodies and Machines* (New York: Routledge, 1992) from which this tidbit comes (p. 105), as well as his analysis of the life and death of individuals in a systematized media world in *Serial Killers* (New York: Routledge, 1998) and in *True Crime: Observations on Violence and Modernity* (New York: Routledge, 2007).

70. *Letters* 1 (30 July 1836), 158.

71. See Frances Ferguson's 'Rape and the Rise of the Novel', *Representations* 20 (1987), 88–112; Ferguson brilliantly reads the skewered print invoking Clarissa's disordered handwriting as a formal means by which Richardson carves out the novel as a psychological genre. My larger point about realism is straight Tucker; see Irene Tucker, *A Probable State: The Novel, the Contract, and the Jews* (Chicago: Chicago University Press, 2000).

72. *Master Humphrey's Clock* (May 1840), 2.71.

73. *Quarterly Review* LIX (1837), 507.
74. Garrett Stewart adeptly dissects Dickens's style in *Dickens and the Trials of Imagination* (Cambridge, Mass.: Harvard University Press, 1974). Stewart trains Dickens readers in such historical fundamentals as Dickens's calculated collisions between high and low language by way of pitting English's everyday short, blunt-sounding words rooted in German and Saxon against their heavy-sounding array of Latin- or Greek-derived synonyms. Stewart's assumption—which still prevails—is, however, that only the rhetorical institutional jargon Dickens attacks represents a contemporary historical outgrowth.
75. 'Winged words, written with running pens, are, in truth, the best adapted to the temper of the times, when the heads of thousands are in a whirl; when time and space are fast hiding their diminished heads, and universal ubiquity, by universal suffrage, is announced to be the order of the day', wrote a reporter about the 'Opening of the Liverpool and Manchester Railroad' in 1830 (*Blackwood's Magazine* 28 (1830), 823). His allusion to Horne Tooke's *EΠEA ΠTEPOENTA [Winged Words], or the Diversions of Purley* (1786), which discusses the abbreviating function of language for the sake of dispatch, demarcates one link between linguistics and transport. Dickens's renowned mastery of shorthand would be another.
76. For a history of cabs, see Trevor May, *Gondolas and Growlers: The History of the London Horse Cab* (Phoenix Mill: Sutton, 1995). As William Long irrefutably details, the scene is chock-a-block with anachronisms. Watermen, for instance, did not have numbers; there could be no cab driver 'No. 924' in 1827 when there were only 100 licenced cabs on the streets. The temporal discombobulation reveals, I believe, that Dickens is encompassing a transformation intended to include a slightly later moment, relevant to his present readers. In 1833 the hackney-coach monopoly ended, and cabs quickly surpassed and replaced hackney coaches, flooding the streets by 1836. See Long, 'Mr Pickwick Lucky to Find a Cab?', 167–70; also 'Streets—Morning' in *Sketches*.
77. On the SPCA see Harriet Ritvo, *The Animal Estate: The English and Other Creatures in the Victorian Age* (Cambridge, Mass.: Harvard University Press, 1987), 127–57; and James Turner, *Reckoning with the Beast: Animals, Pain, and Humanity in the Victorian Mind* (Baltimore: Johns Hopkins University Press, 1980), 39–45.
78. George Orwell, *Dickens, Dali & Others* (New York: Harcourt Brace, 1946); Forster, *The Life of Charles Dickens*; Kate Flint, *Dickens* (Brighton: Harvester, 1986). Mario Praz spat out perhaps the rantiest of the rants against Dickens in this vein in *The Hero in Eclipse in Victorian Fiction*, trans. Angus Davidson (London: Oxford University Press, 1956). Monroe Engel parried with *The Maturity of Dickens* (Cambridge, Mass.: Harvard University Press, 1959).
79. Edmund Wilson, *The Wound and the Bow: Seven Studies in Literature* (London: Methuen, 1961), 26; here is a more recent, fuller reiteration of

the same point:'Charles Dickens was haunted by the idea that the cultivation of a systems-view of the social world—one that analyzed relations of hierarchy and power—was both necessary to the project of realism and potentially highly harmful to individuals, whose critical practices might reify into habits of suspicion that would thwart the bonds of affection that underwrite ideals of family and community' (Amanda Anderson, *The Way We Argue Now* (Princeton: Princeton University Press, 2006), 7).

80. Sally Ledger, *Dickens and the Popular Radical Imagination* (Cambridge: Cambridge University Press, 2007).

CHAPTER 2

On Tragedy's Tracks

1. The strongest theoretical work on nineteenth-century pedestrianism is Celeste Langan's *Romantic Vagrancy:Wordsworth and the Simulation of Freedom* (Cambridge: Cambridge University Press, 1995). Langan argues that, beginning in the early nineteenth century, 'the very behavior [walking] symbolic of political freedom simultaneously connotes social disenfranchisement' (p. 37), and, as a result, historically, 'mobility and agency are on a collision course, constitute a contradiction rather than an identity' (p. 36). See also Langan on 'Mobility Disability' generally in her article of that title in *Public Culture* 13 (2001), 459–84. Langan examines passenger transport in its inequitable economic and political dimensions. Also arguing from Wordsworth, Anne D.Wallace traces *Walking, Literature, and English Culture: The Origins and Use of Peripatetic in the Nineteenth Century* (Oxford: Clarendon Press, 1993);Wallace directly discusses Nell's trek, though not in the larger context of a public transport revolution. Robin Jarvis examines the growth of recreational pedestrian tours in *Romantic Writing and Pedestrian Travel* (Basingstoke: Macmillan, 1997).

2. Wilkie Collins, 'The Last Stage Coachman', *Illuminated Magazine* (August 1843). Dickens details the crumbled aftermath in his 'Uncommercial Traveller' contribution to *All theYear Round* (1 August 1863), later retitled 'An Old Stage-Coaching House'. An old milestone 'looked like a tombstone erected over the grave of the London road' (p. 542).

3. One example, from the United States, of public transport dividing community is the railways' racial segregation in the aftermath of the civil war. I am thinking specifically of the 1896 United States Supreme Court decision in *Plessy v. Ferguson*. In that case, the Court upheld the racist doctrine of separate-but-equal by validating the constitutionality of Louisiana's Separate Car Act for intrastate travel. (The interstate commerce clause of the US Constitution prevented segregation of interstate travel.)

How does it matter that segregating public transport formed the basis for the *Plessy* decision, establishing the doctrine of separate-but-equal,

subsequently extended to other public accommodations, including education? In a searching critical exploration of *Plessy*, Stephen Best brilliantly
calls our attention to the use of counterfactual arguments in the case. As Best
observes, both sides argued the validity of their view by imagining inter-
changing whites and blacks and hypothesizing alternative presents.
Predictably, the result for those setting out from a racist premise was the
reproduction of the necessity for separation: whites in black skins would still
discover the inferiority of the other race and call for their separation.
Conversely, the non-racist counterfactual argument revealed segregation as
based arbitrarily upon the historical accident of birth to a white or black
ancestor: whites finding themselves suddenly in black skins would thus
immediately challenge the indignity actually intended by their segregation.
As Best shows, the racist view suspends historical causation, assigning black
people to an unchanging present. As he also observes, however, both sides'
counterfactual arguments problematically call on empathy, sympathy, and
sentiment, premising their appeals upon the invisible, interior contents of
character against skin. As we have seen, the interchangeability of railway passengers ignores such contents. Thus, at least in this aspect of *Plessy*, the contents of the two sides' counterfactual arguments would seem to be
contradicted by the formal relations set up by the railways. (This may partly
explain why the Louisville and Nashville Railway Company cooperated in
arranging Plessy's arrest and opposed segregation, at least in the shape of
separate railway cars, as presenting unnecessary cost.) One implication is
perhaps that the sort of interchangeability produced by a passenger network
represented that which needed to be overcome for a racist outcome. *Plessy*'s
later reversal by the decision in *Brown v. Board of Education* also only represented the triumph of the opposing counterfactual; it outlawed segregation
in education, precisely where the equality of the contents of character is at
stake. Such considerations point, I hope, toward how one might foreground
the structuring of a transit networked community in examining its use as
site for segregation and exclusion. See Stephen M. Best's chapter 'Counter-
factuals, Causation, and the Tenses of "Separate but Equal"' in *The Fugitive's
Properties: Law and the Poetics of Possession* (Chicago: University of Chicago
Press, 2004). On passenger transport's brutal and divisive British colonialist
and class history, see Joanna Guldi's *Roads to Power: Britain Invents the
Infrastructure State* (Cambridge, Mass.: Harvard University Press, 2012).

4. 'The soul of tragedy is the plot', Aristotle famously writes in the *Poetics*.
He is explaining that the depiction of human suffering alone is not tragedy's linchpin, and a plot in which Nell died, say, in the single gentleman's
arms would be sad but not necessarily tragic. In a classically Aristotelian
tragedy—on offer in this novel—the forces that hold people together in
community also turn out to produce the conditions tearing them apart
(which is a pity) and expose the vulnerability inhering in those same communal relations (which is terrifying).

5. The most important previous essay to broach the story's framing by Master Humphrey's clock in its opening and St Paul's clock at its end is Philip Rogers's 'The Dynamics of Time in *The Old Curiosity Shop*', *Nineteenth-Century Fiction* 28 (1973), 127–44.

6. Steven Marcus, *Dickens from Pickwick to Dombey* (New York: Simon and Schuster, 1965), 131.

7. *Master Humphrey's Clock* in 20 monthly parts (London: Chapman & Hall), 2.74. Further citations to *Master Humphrey's Clock*, including to *The Old Curiosity Shop*, are given parenthetically and refer to the original serial by monthly number and page. (Pagination was restarted after the sixth monthly number, which completed Volume One of *Master Humphrey's Clock*.) It is much better known that *Clock* came out weekly. Both the weekly and the monthly issue, however (resulting in two novels—*The Old Curiosity Shop* and *Barnaby Rudge*—each also measuring as roughly a half-sized triple-decker at ten-and-a-half parts), illuminate these stories' structuring.

8. G. K. Chesterton, *Appreciations and Criticisms of the Works of Charles Dickens* (London: Dent, 1911), 52–3.

9. Thomas Hood, review of '*Master Humphrey's Clock*. By 'Boz'. Vol. I. Chapman & Hall', *The Athenaeum* (7 November 1840), 887.

10. A strikingly similar imaginative perspective structures Dickens's 'Familiar Epistle From a Parent to a Child Aged Two Years and Two Months', *Bentley's Magazine* (February 1839), in which Dickens was handing over the magazine's editorship to Ainsworth (which makes it especially relevant to Dickens's resumption of magazine editorship with *Clock*). In this short piece, Dickens recalls observing a mail guard riding a train, remembers stage coaching and imagines its displacement in the future, and ties it all to the magazine's conducting the readers as passengers, signing off as 'the old coachman, Boz'. Dickens also satirizes the notion that industrial railway drivers were conspiring against agrarian coachmen in the *Morning Chronicle* (9 March 1844), 64.

11. Paul Virilio, *Speed and Politics: An Essay on Dromology*, trans. Mark Polizzotti (New York: Semiotext(e), 1986), 30. See also Michel de Certeau's chapter 'Railway Navigation and Incarceration' for an influential Foucauldian meditation on the panopticism of railways, especially relevant here for its juxtaposition with his chapter on 'Walking in the City' as offering an everyday tactic for preserving individual autonomy, in *The Practice of Everyday Life*, trans. Steven Rendall (Berkeley: University of California Press, 1984).

12. On a parallel moment in George Eliot, see Evan Cory Horowitz, 'George Eliot: The Conservative', *Victorian Studies* 49 (2006), 7–32.

13. On the history of standardizing time, see Derek Howse, *Greenwich Time and the Discovery of Longitude* (Oxford: Oxford University Press, 1980). Many railway histories provide some brief accounts of the formation of

standard time; Wolfgang Schivelbusch goes further in *The Railway Journey: The Industrialization of Time and Space in the 19th Century* (Berkeley: University of California Press, 1986). David Landes's *Revolution in Time: Clocks and the Making of the Modern World* (Cambridge, Mass.: Belknap Harvard, 2000) provides the authoritative horological history of time's keeping and its makings; his pointedly titled chapter 'My Time is Your Time' is especially relevant here. See also Gerhard Dohrn-van Rossum, *History of the Hour: Clocks and Modern Temporal Orders*, trans. Thomas Dunlap (Chicago: University of Chicago Press, 1996). In *Technics and Civilization* (New York: Harcourt, 1934), Lewis Mumford famously suggested that 'the clock, not the steam engine, is the key-machine of the modern industrial age' (p. 14). As part of Mumford's incontrovertible deepening and widening of the past historical forces that led to the embrace of mechanization, he returns to the collective time produced in monasteries. Where Mumford provocatively looks before the period under examination here, Peter Galison expertly recounts the history of the era that follows, looking at the turn of the twentieth century in *Einstein's Clocks, Poincaré's Maps: Empires of Time* (New York: Norton, 2003).

I am focusing on only one part of the history of standard time and, most obviously, ignoring the historical effort to map longitude (the knowledge of the measure of east–west distances on the globe that enables us to relate its turning to the sun and therefore to the day) along with the telegraph's creation of near-instant communication across space (which followed up the portable, super-accurate clocks in enabling the accurate triangulation of space in time required to map longitude). I am also not treating, though Dickens does treat it with great acuity, the shift that occurred between time tolled by bells (creative of collectivity in coming to its auditors) and time being told by public clocks and private watches (creative of collectivity through their individual synchronization). Much could be written on just this in relation to *The Old Curiosity Shop*.

14. See, for instance, a letter of 1 September 1795 reprinted in Edmund Vale's *The Mail-Coach Men of the Late Eighteenth Century* (Newton Abbot: David & Charles, 1967), 136–7.

15. *Illustrated London News* (May 1842), 16.

16. 'Early Great Western Timetable, 1840s', in Science and Society Picture Library (National Railway Museum), Image Reference, 10290335.

17. Chesterton, *Appreciations and Criticisms of the Works of Charles Dickens*, 53.

18. Peter K. Garrett, *The Victorian Multiplot Novel: Studies in Dialogical Form* (New Haven: Yale University Press, 1980), 8, 10. Helena Michie investigates how the simultaneity of plot threads gets produced at the sentence-level; she juxtaposes both realist multiplots (where the sentence may cue the reader to think of untold, ongoing plot threads) and writing histori-

cally about it in 'Victorian(ist) "Whiles" and the Tenses of Historicism', *Narrative* 17 (2009), 274–90.

19. Because simultaneity must be reconstructed, no such thing as a narrative 'linear time' exists. As far as I can tell, in literary studies 'linear time' is paraded out as that which is endlessly being contradicted or refuted, a kind of stooge, useful for setting off supposedly more complex temporalities. It is often someone else's simplistic idea of causation. 'Linear time' has existed strongly in this mirage form, and it perhaps has hampered our critical analyses, especially of the temporalities actually produced by the clocks and calendars that supposedly abet it. Steven Jay Gould traces the metaphor's long history in *Time's Arrow, Time's Cycle: Myth and Metaphor in the Discovery of Geological Time* (Cambridge, Mass.: Harvard University Press, 1987), but the phrase 'linear time' is a twentieth-century cultural phenomenon, with usage dramatically increasing after 1960 (Google Ngram search, 'linear time').

20. Philip S. Bagwell, *The Transport Revolution* (London: Routledge, 1974), 13. On the history of canals, see Charles Hadfield, *British Canals: An Illustrated History* (London: Phoenix, 1950) and *The Canal Age* (Newton Abbot: David & Charles, 1968); Jean Olivia Lindsay, *The Canals of Scotland* (Newton Abbot: David & Charles, 1968); and L. T. C. Rolt, *The Inland Waterways of England* (London: Unwin, 1950). Scant literary critical attention has been paid to the rise of the canals; Markman Ellis has a smart chapter on them in relation to the sentimental novel and commerce in *The Politics of Sensibility: Race, Gender and Commerce in the Sentimental Novel* (Cambridge: Cambridge University Press, 1996).

21. Ruskin famously declared that Nell was butchered for the market; Adorno saw her clinging to things in a pre-bourgeois fashion. On Nell as 'The Ideal Girl in Industrial England', see the chapter of that title in Catherine Robson's *Men in Wonderland: The Lost Girlhood of the Victorian Gentleman* (Princeton: Princeton University Press, 2001), 46–93; Sue Zemka discusses class in 'From the Punchmen to Pugin's Gothics: The Broad Road to a Sentimental Death in *The Old Curiosity Shop*', *Nineteenth-Century Literature* 48 (1993), 291–309. Building on E. P. Thompson's important essay 'Time, Work-Discipline, and Industrial Capitalism', N. N. Feltes argues Dickens depicts time as commodified in *Shop*'s industrial scenes in 'To Saunter, to Hurry: Dickens, Time, and Industrial Capitalism', *Victorian Studies* 20 (1977), 245–67.

On the economics of transport history, see H. J. Dyos and D. H. Aldcroft's *British Transport: An Economic Survey from the Seventeenth Century to the Twentieth* (Leicester: Leicester University Press, 1969). Even at the time it was well understood that transit corridors systematically obliterate views of poverty, as Friedrich Engels noted of the layout of Manchester in *The Condition of the Working Class in England* in 1844. (Steven Marcus explains Engels's relevance to Dickens's writing in *Engels, Manchester, and the Working Class* (New York: Norton, 1974).) The effect and its study continues; e.g.

Xina M. Tadiar thoughtfully describes the building of highway flyovers in Manila in 'Manila's New Metropolitan Form', *differences* 5 (1993), 154–79.

22. Harry Hanson, *The Canal Boatmen, 1760–1914* (Gloucester: Sutton, 1984).

23. Ibid. 71.

24. See Walter Dexter's 'On the Track of Little Nell', in *The England of Dickens* (Philadelphia: Lippincott, [1925]). For a map charting Nell's route from London through Birmingham to Tong see Albert A. Hopkins and Newbury Frost Read, *A Dickens Atlas* (New York: Hatton Garden, 1923), Plate XXII. Far from wandering all about England, Nell makes a one-way, north–west journey.

25. *Letters* 2 (?3 April 1840), 50. Dickens was going to visit his brother, Alfred Lambert Dickens, who was working on the Birmingham and Derby railway as a surveyor and civil engineer. He was living in Tamworth, just north-east of Birmingham; see *Letters* 2 (4 March 1840), 223. On Dickens's 1838 trip, which provided material for the novel (as he told Forster), Dickens records how he was forced through Birmingham and Wolverton, north of what would be a direct north–west route from Stratford to Shrewsbury on a lesser road: 'We remained at Stratford all night, and found to our unspeakable dismay that father's plan of proceeding by Bridgenorth was impracticable as there were no coaches. So we were compelled to come here by way of Birmingham and Wolverhampton, starting at eight o'Clock through a cold wet fog, and travelling when the day had cleared up, through miles of cinder-paths and blazing furnaces and roaring steam engines, and such a mass of dirt gloom and misery as I never before witnessed. We ... arrived at half past four, and are now going off in a post-chaise to Llangollen—30 miles— where we shall remain to-night, and where the Bangor Mail will take us up tomorrow.... My side has been very bad since I left home ... I suffered such an ... pain all night at Stratford ... If I had not got better I should have turned back to Birmingham, and come straight home by the railroad. As it is, I hope I shall make out the trip' (*Letters* 1 (1 November 1838), 447–8). As it was, Forster met them in Liverpool having travelled up by rail (on a route opened seven weeks earlier), and they return together by rail: 'We shall leave here at 3 o'Clock on Thursday *Morning*, and shall be at the Euston Square Station of the Birmingham Railway—I should think about two on Thursday afternoon' (*Letters* 1 (5 November 1838), 449–50). The trip as a whole nicely foregrounds the ongoing standardizing of space along with various intersecting modes of mobility.

26. Philip Rogers, 'The Dynamics of Time in *The Old Curiosity Shop*', *Nineteenth-Century Fiction* 28 (1973), 127–44.

27. Andrew Sanders leads the pack who believe that what crucially separates today's readers from the Victorians is having lost a historical context of familiarity with child death. Sanders's *Charles Dickens Resurrectionist* (New

York: St. Martin, 1982) begins by announcing, 'Little Nell's death is virtu-ally the goal of the progress traced in *The Old Curiosity Shop*' (p. 1). Hilary Schor is more helpfully ambiguous in suggesting that 'The central activity of any reader of *The Old Curiosity Shop* is watching Little Nell walk herself to death' (*Dickens and the Daughter of the House* (Cambridge: Cambridge University Press, 1999), 32).

28. John Bowen, *Other Dickens: Pickwick to Chuzzlewit* (Oxford: Oxford University Press, 2000); Marcus, *Dickens from Pickwick to Dombey*; Gabriel Pearson, 'The Old Curiosity Shop', in John Gross and Gabriel Pearson (eds.), *Dickens and the Twentieth Century* (London: Routledge, 1962); Garrett Stewart's *Trials of the Imagination* (Cambridge, Mass.: Harvard University Press, 1974) in particular makes much of Swiveller's branching off.

29. Jerome Meckier, 'Suspense in *The Old Curiosity Shop*: Dickens' Contrapuntal Artistry', *The Journal of Narrative Technique* 2 (1972), 199–207.

30. 'Come and lie down', Katie Dickens said to her father when he began to suffer a brain hemorrhage on the ninth of June 1870. (It was five years to the day after his train crashed at Staplehurst.) She reports that he replied with his last words: 'Yes, on the ground.' One may perhaps wonder, how-ever, if she could have distinguished 'on the ground' from 'un-der-ground', or if his words might have carried both senses. In *Little Dorrit*, Dickens describes the newly dead as being 'removed by an untraversable distance from the teeming earth and all that it contains, though soon to lie in it' (16.492).

31. The quarrel over sympathizing or not with Nell began immediately upon publication: it is fully as old as the novel and aging right along with it. Setting up, however, on either side of this divide over Nell sharply dimin-ishes the novel and, historically, has thus contributed to the critical momentum of its depreciation. See Priscilla Schlicke and Paul Schlicke (eds.), *The Old Curiosity Shop: An Annotated Bibliography* (New York: Garland, 1988) for a historical summary of the novel's reception.

32. Simultaneity with the dead is the subject of William Wordsworth's poem 'We are seven', one of Dickens's favorites. Dickens plays on the paradox and the poem (reversing its roles) when Nell meets an old woman at her long-deceased husband's graveside. Critics have not overlooked the allu-sion to Wordsworth; see, for notable instance, Angus Wilson, *The World of Charles Dickens* (New York: Viking, 1970), 144.

33. Audrey Jaffe, *Vanishing Points: Dickens, Narrative, and the Subject of Omniscience* (Berkeley: University of California Press, 1991), 7, 4.

34. Sylvère Monod, *Dickens the Novelist* (Norman: University of Oklahoma Press, 1968), 171–5.

35. *Letters* 2 (6 November 1840), 147.

36. Elizabeth M. Brennen, 'Little Nell on Stage, 1840–1841', in Charles Dickens, *The Old Curiosity Shop*, ed. Elizabeth M. Brennen (Oxford: Clarendon Press, 1997), 634.

37. Forster MS, *The Old Curiosity Shop*, National Arts Library. The manuscript includes the ending in *Clock* as well.

38. Raymond Williams, *The Country and the City* (New York: Oxford University Press, 1973), ch. 15, 'People of the City'. The famous and brilliant passage, which I am suggesting needs revision and which couples its insight too strictly solely to the city, is Williams's contention that 'as we stand and look back at a Dickens novel the general movement we remember—the characteristic movement—is a hurrying seemingly random passing of men and women, each heard in some fixed phrase...a way of seeing men and women that belongs to the street. There is at first an absence of ordinary connection and development. These men and women do not so much relate as pass each other and then sometimes collide.... But then as the action develops, unknown and unacknowledged relationships, profound and decisive connections, definite and committing recognitions and avowals are as it were forced into consciousness' (p. 155).

CHAPTER 3

International Connections

1. Charles Dickens, *Little Dorrit*, 20 numbers in 19 monthly parts (London: Bradbury & Evans, 1855–7), 7.382. All citations to *Dorrit* are to the monthly number and page.

2. Lionel Trilling, 'Little Dorrit', *The Kenyon Review* 15 (1953), 577–90; John Lucas, *The Melancholy Man: A Study of Dickens's Novels*, 2nd edn. (Brighton: Harvester Press, 1980); J. Hillis Miller, *Charles Dickens: The World of His Novels* (Cambridge, Mass.: Harvard University Press, 1958); Hilary Schor, 'Novels of the 1850s', in John O. Jordan (ed.), *The Cambridge Companion to Charles Dickens* (Cambridge: Cambridge University Press, 2001), 70; see Hilary Schor's essay and her discussion of *Dorrit* in *Dickens and the Daughter of the House* (Cambridge: Cambridge University Press, 1999) for expert elucidations of this novel's plot.

3. John Carey, *The Violent Effigy: A Study of Dickens' Imagination* (London: Faber and Faber, 1973), 113–17.

4. See William Burgan, '*Little Dorrit* in Italy', *Nineteenth Century Fiction* 29 (1975), 393–411; Nancy Aycock Metz, '*Little Dorrit's* London: Babylon Revisited', *Victorian Studies* 33 (1990), 465–86; Tore Rem, '*Little Dorrit, Pictures from Italy*, and John Bull', in Anny Sadrin (ed.), *Dickens, Europe and the New Worlds* (New York: St. Martin's, 1999), 131–45. Amanda Anderson correctly realized that no synthesis can hold together the novel's cosmopolitan outlook and its hermeneutics of suspicion, but rather than questioning the monolithic cast she then forces onto the novel in requiring such a synthesis from it, Anderson dismisses Dickens as 'reductively moralizing' ('Cosmopolitanism in Different Voices: Charles Dickens's *Little Dorrit* and the Hermeneutics of Suspicion', in *The Powers of Distance:*

Cosmopolitanism and the Cultivation of Detachment (Princeton: Princeton University Press, 2001), 89).

The list of 'Houses Where Dickens Lived and Worked' offers a powerful condensation of Dickens's turn toward an international life beginning in 1844 (Albert A. Hopkins and Newbury Frost Read's *A Dickens Atlas* (New York: Hatton Garden, 1923), 5); while the plot of *Little Dorrit* almost shockingly overlaps with the useful map of Dickens's travels in Europe in Paul Schlicke (ed.), *Oxford Reader's Companion to Dickens* (Oxford: Oxford University Press, 1999).

5. Plan for No. XVI, Forster collection, Victoria and Albert Museum. Dickens's number plans are widely reprinted; *Dickens' Working Notes for His Novels* (Chicago: University of Chicago Press, 1987), edited and introduced by Harry Stone, offers photographs as well as typescripts of the number plans.

6. Michel Foucault, *Discipline and Punish: The Birth of the Prison*, trans. Alan Sheridan (New York: Vintage, 1979); D. A. Miller, *The Novel and the Police* (Berkeley: University of California Press, 1988).

7. Lucas, *The Melancholy Man*, 248.

8. On the temporality of the pictorial and the narratorial in *Little Dorrit*, see Martin Meisel, *Realizations: Narrative, Pictorial, and Theatrical Arts in Nineteenth-Century England* (Princeton: Princeton University Press, 1983), 319–20.

9. Peter Brooks, *Reading for the Plot: Design and Intention in Narrative* (Cambridge, Mass.: Harvard University Press, 1984), 23.

10. My image of time flapping on the mast comes from Virginia Woolf's *Mrs Dalloway*, a novel with some overlapping concerns to *Little Dorrit* and one that is often mistakenly read (e.g. by Stephen Kern) as evaluatively pitting an inner psychological time, gendered female, against a disruptive official and masculine—Big Ben—standard time. In fact, Woolf's narrator brings into play the synchronous connectivity produced by standardizing time, and when she tunnels furthest into her characters' minds she also finds there an androgynous, simultaneously interconnected 'All'.

11. For an introduction to the history of steam ships, see K. T. Rowland, *Steam at Sea: A History of Steam Navigation* (Newton Abbot: David & Charles, 1970); on Brunel's extraordinary career, see L. T. C. Rolt, *Isambard Kingdom Brunel, a Biography* (New York: St. Martin's, 1959). P. J. G. Ransom is particularly useful in helping to show the interleaving co-development of inland passenger transport alongside the trans-oceanic (*The Archeology of the Transport Revolution, 1750–1850* (Tadworth: World's Work Ltd., 1984)). Philip Bagwell details the explosion of national steamship services after 1812; the island nation, especially before the rail system was fully established, developed a high-speed, coastal network all around it interlinking all its ports. In an important early test, travelling from Glasgow to London, the PS *Thames* proved in 1815 that the steam boats could effectively plow directly against

even heavy winds. See Philip S. Bagwell, *The Transport Revolution* (London: Routledge, 1974), ch. 3.

12. Charles Dickens, *Pictures from Italy* (London: Bradbury & Evans, 1846), 186. Insofar as Dickens's characters slalom purposefully around on passenger transport networks, they contrast with touristic travel—via those networks—for travel's sake, which Dickens exhibits in the Dorrits' outdated Grand Tour. For tourists, paradoxically, 'shared values [do not] create solidarity within the international community of tourists but hostility, as each wished the other tourists were not there', making for 'a cultural consensus that creates hostility rather than community', as Jonathan Culler observes in his seminal essay 'The Semiotics of Tourism', *Framing the Sign* (Oxford: Basil Blackwell, 1988), 158. A similar phenomenon happens for different reasons in overcrowded corridors of public transit. On the topic of tourist travel, see James Buzard, *The Beaten Track: European Tourism, Literature, and the Ways to Culture, 1800–1918* (Oxford: Clarendon Press, 1993) and Ali Behdad's post-colonial rethinking of nineteenth-century tourism in *Belated Travelers: Orientalism in the Age of Colonial Dissolution* (Durham, NC: Duke University Press, 1994).

13. 'The Streets by Morning', in *Sketches by 'Boz', Illustrative of Every-day Life, and Every-day People*, Second Series (London: Macrone, 1837), 10; first published in the *Evening Chronicle*, 21 July 1835.

14. *Letters* 7 (12 November 1854), 464.

15. Charles Dickens, *Dealings with the Firm of Dombey and Son, Wholesale, Retail, and for Exportation*, 20 numbers in 19 monthly parts (London: Bradbury & Evans, 1846–8), 5.155.

16. Derek Howse, *Greenwich Time and the Discovery of Longitude* (Oxford: Oxford University Press, 1980). See also David Landes's *Revolution in Time* (Cambridge, Mass.: Belknap Harvard, 2000). The impetuses and developments that led to standard time were many and complex. I am starkly focusing on one.

17. 'Out of the Season', *Household Words* (28 June 1856), 553. The counterpart to this sketch, published the year before, is called 'Out of Town' (*Household Words*, 29 September 1855). In it, Dickens describes the Great Hotel Pavilion in Folkestone, renaming the town itself in the sketch Pavilionstone after the hotel. Folkestone had become the busy, new departure point for the Continent, where 'you shall find all the nations of the earth': 'Couriers you shall see by hundreds', and 'more luggage in a morning than, fifty years ago, all Europe saw in a week' (p. 195); 'if you want to live a life of luggage, or to see it lived ... come to Pavilionstone' (p. 196).

18. Placard displayed at Greenwich Observatory (2005); also Howse, *Greenwich Time*, 113; Kristen Lippincott (ed.), *The Story of Time* (London: Merrell Holberton, 1999), 146.

19. Stuart Sherman, *Telling Time: Clocks, Diaries, and English Diurnal Form, 1660–1785* (Chicago: University of Chicago Press, 1996).

20. John North, *God's Clockmaker: Richard of Wallingford and the Invention of Time* (London: Hambledon, 2005), 4.

21. Sandford Fleming, *Uniform Non-Local Time (Terrestrial Time)* (Ottawa, 1876). Coordinated Universal Time replaces GMT in the twentieth century. It is kept by a set of international atomic clocks (keyed to the vibrations of cesium atoms), whose mean generates the time. Next in time for time may be a Universal Time kept with PARCS, a Primary Atomic Reference Clock in Space, which will be connected to the Global Positioning System (GPS) that already provides orbiting reference clocks, whose relations register spatial positioning on the globe. With the coming of PARCS, global time officially will have left the planet. GPS time and Universal time are not currently in synch.

22. *Proceedings of the Canadian Institute*, Third Series, Vol. 3, 1884–5 (Toronto, 1886), 75.

23. In the 1960s and 1970s, simultaneity's edges expanded around 'Spaceship Earth', most famously captured in the self-observational photograph called 'The Blue Marble' (1972). History had by that point fully ironized Hazlitt's assumption in his essay 'On a Sun-Dial' that 'We might as well make a voyage to the moon as think of stealing a march upon Time' (*The Complete Works of William Hazlitt*, ed. P. P. Howe (London: Dent, 1933), 17.242). Instead, as Paul Virilio speculates, the implied future would seem to be that 'all of Earth's inhabitants may well wind up thinking of themselves more as *contemporaries* than as *citizens*' (*The Art of the Motor*, trans Julie Rose (Minneapolis: University of Minnesota Press, 1995), 36). Virilio does not think that this would necessarily be a good thing.

24. Benedict Anderson, *Imagined Communities: Reflections on the Origin and Spread of Nationalism*, rev. edn. (London: Verso 2006), 26. At the end of his study, Anderson offers a glimpse of 'the cosmic clocking which had made intelligible our synchronic transoceanic pairings' (p. 194). My argument is that the international passenger transport system, which Anderson overlooks, helped give rise to that cosmic clocking productive of (in his nice phrase) 'our synchronic transoceanic pairings'.

25. Ian Duncan, *Modern Romance and Transformations of the Novel: The Gothic, Scott, Dickens* (Cambridge: Cambridge University Press, 1992); Katie Trumpener, *Bardic Nationalism: The Romantic Novel and the British Empire* (Princeton: Princeton University Press, 1997); Priya Joshi, *In Another Country: Colonialism, Culture, and the English Novel in India* (New York: Columbia, 2002); James Buzard, *Disorienting Fiction: The Autoethnographic Work of Nineteenth-Century British Novels* (Princeton: Princeton University Press, 2005). As *Pickwick*'s deflationary rejection of travel literature especially helps make clear, public transportation shrinks distance and time, making for precisely the opposite scenario than that which Buzard proposes: 'here' is extending to include 'there'. And while in anthropology, as Johannes Fabian insightfully saw in *Time and the Other: How Anthropology Makes Its*

Object (New York: Columbia, 1983), anthropologists were interacting with others in the field but then suppressing their coevality in discoursing about those others, that discourse is—meaningfully—not the same as multiplotted fiction.

Recently, Franco Moretti's credo has been that separate national histories of the novel obfuscate the genre's global history. Hence he has embarked on ambitious collective transnational histories and theorizations of the novel. Without scanting this laudable aim to overcome the novel's yoke to nation by rediscovering it as 'a great anthropological force . . . the first truly planetary form', one might note that doing so is in danger of mistaking at the outset the novel's ascription to a nation as nationalistic instead of international (Franco Moretti (ed.), *The Novel* (Princeton: Princeton University Press, 2006), p. ix).

26. Ian Baucom, *Out of Place: Englishness, Empire, and the Locations of Identity* (Princeton: Princeton University Press, 1999).

27. Anderson, *Imagined Communities*, 56, 44.

28. During the writing of *Dorrit*, Dickens first became involved with the translation of his works. His correspondence (in French) with the publishing house Hachette begins in December 1855; dinner parties with the publisher and translators in Paris follow as well. He would reread the l'eau/low joke in French in Hachette's authorized French translation of *Nickleby* just a few months after replaying it in *Dorrit* (*Letters* 8 (3 April 1857), 308).

29. Irene Tucker, *A Probable State: The Novel, The Contract, and the Jews* (Chicago: University of Chicago Press, 2000), 9.

30. At some unrecoverable point, the toast 'To all friend round St Paul's' began to include 'and may the circle have no bounds'.

31. *Psychology Today* 1 (1967). Microsoft reconfirmed the phenomenon in 2008 by finding that in their global instant messaging system people were connected to others through an average of 6.6 steps (BBC News, 2008/08/03,http://news.bbc.co.uk/go/pr/fr/-/2/hi/technology/7539329.stm). The widespread use of GPS is beginning to allow for the detecting of mobility networks (instead of just communication networks) as well as GPS art.

32. John Forster, *The Life of Charles Dickens*, ed. J. W. T. Ley (Cecil Palmer, 1928), 76.

33. Sylvia Manning, 'Social Criticism and Atextual Subversion in Little Dorrit', *Dickens Studies Annual* 20 (1991), 140. Garrett Stewart, by salutary contrast with Manning however, has recently deep-scanned the way the mad mother and marriage plot troubles this novel in *Novel Violence: A Narratography of Victorian Fiction* (Chicago: University of Chicago Press, 2009).

34. Hilary M. Schor, 'Dickens and Plot', in John Bowen and Robert L. Patten (eds.), *Palgrave Advances in Charles Dickens Studies* (Basingstoke: Palgrave, 2006), 101–2. Janice Carlisle grouches that Miss Wade has issued 'empty

promises' in 'Little Dorrit: Necessary Fictions', Studies in the Novel 7 (1975), 205; Amanda Anderson reiterates: 'The narrator's own earlier vexed identification with Miss Wade becomes immaterial as we discover that her oracular stance in chapter 2 was deceptive, and that she had specific personal reasons for being in the company of the Meagles' (The Powers of Distance: Cosmopolitanism and the Cultivation of Detachment (Princeton: Princeton University Press, 2001), 82); Peter Garrett deduces a complete anti-climax: 'the real secret is an absence of determinate, objective meaning which the elaborate machinery of the mystery plot can no longer quite hide, a void in which it spins its wheels, going nowhere' (The Victorian Multiplot Novel: Studies in Dialogical Form (New Haven: Yale University Press, 1980), 75).

35. Neil Forsyth and Peter Garrett do quote the rest of Wade's speech but do not pursue its implications; Forsyth in 'Wonderful Chains: Dickens and Coincidence', Modern Philology 83 (1985), 162; and Garrett in Gothic Reflections: Narrative Force in Nineteenth-Century Fiction (Ithaca: Cornell University Press, 2003), 164–5. Sylvia Manning quotes the speech in full, and she glimpses in the second part of Miss Wade's speech its pointer to 'when Rigaud attaches himself to Gowan' but then dismisses it—'nothing happens' (pp. 140–1).

36. In the third number plan, a query—'Miss Wade in the prison? . . . Her father?'—suggests Dickens may have toyed with having Miss Wade be Rigaud's daughter.

37. Letters 8 (?9 February 1857), 279–80.

38. John Frow, 'Voice and Register in Little Dorrit', Comparative Literature 33 (1981), 264–5. Where Frow goes on to pursue the significance of the Gothic prison plots, I am pursuing the significance of the journeying plots. See also 'Dickens and the Sense of Time', in John Henry Raleigh's Time, Place, and Idea: Essays on the Novel (Carbondale: Southern Illinois University Press, 1968), esp. 133.

39. Iain Crawford, ' "Machinery in Motion": Time in Little Dorrit', The Dickensian 84 (1988), 30–41.

40. In the Preface, Dickens presents himself as stepping out from behind the curtain of omniscient narration to expose the limited status of his own individual authorial knowledge. 'I . . . have held [this story's] various threads', he declares, but then continues: 'I did not know myself, until the sixth of this present month' if the edifice of the Marshalsea prison still remains (19&20, pp. v–vi). He then recounts a visit to the street-cum-prison. There Dickens encounters a young boy, who calls attention to undisclosed webs of interconnections that Dickens, as an individual like the reader, cannot access. Standing in the street-prison, he is thus unable to know all its 'crowding ghosts' (19&20, p. vii). Andrew Miller inducts us into the ghostly side of this realist phenomenon: the depiction of a life led as imagined against lives unled; see his work on this optative mode of narrating, begun

in *The Burdens of Perfection: On Ethics and Reading in Nineteenth-Century British Literature* (Ithaca: Cornell University Press, 2008).

41. My discussion of Affery's dreams is indebted to Jay F. Rosenberg's *Accessing Kant: A Relaxed Introduction to the Critique of Pure Reason* (Oxford: Clarendon Press, 2005), esp. 86–7. The conclusion of this chapter and its close readings are deeply indebted to personal conversations with Adam Parker.

AFTERWORD

1. Ruth Livesey, 'Communicating with *Jane Eyre*: Stagecoach, Mail, and the Tory Nation' (forthcoming in *Victorian Studies*).
2. Nikolai Gogol, *Dead Souls*, trans. Richard Pevear and Larissa Volokhonsky (New York: Vintage, 1997), 253.
3. Larry Duffy takes Zola's novel as the starting point for his investigation of naturalist fiction in relation to transport history in *Le Grand Transit Moderne: Mobility, Modernity, and French Naturalist Fiction* (Amsterdam: Rodopi, 2005).

Index

Page numbers in italics refer to illustrations.